The Weight of the Social Economy

An International Perspective

P.I.E. Peter Lang

Bruxelles · Bern · Berlin · Frankfurt am Main · New York · Oxford · Wien

CIRIEC
Marie J. Bouchard and Damien Rousselière (eds.)

The Weight of the
Social Economy

An International Perspective

Social Economy & Public Economy
No.6

CIRIEC activities, publications and researches are realized with the support of the Belgian Federal Government – Scientific Policy and with the support of the Government of the Belgian French Speaking Community – Scientific Research.

This publication has been peer-reviewed.

© P.I.E. PETER LANG S.A.
    Éditions scientifiques internationales
    Brussels, 2015
    1 avenue Maurice, B-1050 Bruxelles, Belgique
    info@peterlang.com ; www.peterlang.com

ISSN 2030-3408
ISBN 978-2-87574-287-2
eISBN 978-3-0352-6545-3
D/2015/5678/51

Printed in Germany

CIP available from the British Library, GB and the Library of Congress, USA.
Bibliographic information published by "Die Deutsche Nationalbibliothek"
"Die Deutsche Nationalbibliothek" lists this publication in the "Deutsche Nationalbibliografie"; detailed bibliographic data is available in the Internet at http://dnb.d-nb.de.

Contents

8

Acknowledgements

The present book is the result of the CIRIEC International working group "The Weight, Size and Scope of the Social Economy." At the closing of *The Weight of the Social Economy. An International Perspective*, we would like to thank all the members of the working group for their collaboration and the quality of our conversations. We thank Édith Archambault and Université de Paris 1 Panthéon-Sorbonne for hosting the founding meeting of this group, as well as participants of that meeting (Gian Nicola Francesconi, International Food Policy Research Institute (Senegal); Karine Latulippe, Institut de la statistique du Québec (Canada)), and the contributors to this book for their valuable contributions. This book could not have been produced without the support of CIRIEC International staff, in particular Christine Dussart and Carmela De Cicco. We also wish to acknowledge Rafael Chaves, Chair of the CIRIEC scientific commission "Social and Cooperative Economy" until 2015, and the editors of the Social Economy & Public Economy series of P.I.E. Peter Lang: Benoît Lévesque (Université du Québec à Montréal, Canada) and Bernard Thiry (Université de Liège and general director of CIRIEC International, Belgium). Finally, we thank our translator and editor Cathleen Poehler for her professionalism and insightful suggestions throughout the process of finalizing this book.

Marie J. Bouchard and Damien Rousselière

Introduction

The Weight, Size and Scope of the Social Economy

Marie J. BOUCHARD

Full professor, Université du Québec à Montréal, Canada

Damien ROUSSELIÈRE

Full professor, AGROCAMPUS OUEST, France

1. Objectives

There is a growing interest in statistics about the social economy. Throughout the world, the social economy plays an important role in job creation and retention, social cohesion, social innovation, rural and regional development, and environmental protection. In a number of countries, significant work has been done by national statistical agencies and by researchers (academic and institutional) to gather national and international data. Results tend to confirm the ability of the social economy to contribute to balancing economies, mainly by serving as an anti-cyclical force in the face of economic crises (Stiglitz, 2009; European Parliament and European Commission, 2014). The social economy is seen as having a strong potential in addressing global challenges in a renewed economy (Utting, van Dijk and Matheï, 2014).

Still, many countries and regions lack information about the social economy on their territory. Nor is there a harmonized body of data on the social economy at the global level. Moreover, a body of knowledge for promoting the importance of the social economy in the public realm has not been compiled to date. The purpose of measuring a phenomenon is to ensure its social and political recognition, namely to justify government support (Anheier, Knapp and Salamon, 1993). Here, public authorities and key players in the system have formulated requests to measure and evaluate the contribution of the social economy. Demands are made to identify the size and scope of this type of economy, especially during a time when other dominant systems seem to fail in their response to many social and economic needs (Stiglitz, Sen and Fitoussi, 2009). However, as noted by the Research Working Group of the 2014 Social Economy conference in Rome:[1]

> Systematic data collection seems to be a common problem, and coordination between various statistical offices is required. Recurrent issues relate to the quality of statistics, and the absence of quality data in many cases. This seemed to be related to the lack of a clear definition in order to define the population within much larger datasets. The social economy has a large range of (often contested) meanings, and so this raises quantitative challenges (Roy, 2014).

CIRIEC (Centre International de Recherches et d'Information sur l'Économie Publique, Sociale et Coopérative / International Centre of Research and Information on the Public, Social and Cooperative Economy) decided to launch an international project on the production of statistics on the social economy.[2] The goal was to bring together contributions that are taking stock of knowledge on the production of statistics on the social economy. Called *The Weight, Size and Scope of the Social Economy. An International Perspective on the Production of Statistics for the Social Economy*, the project aimed to spur interest in the social economy and to get national statistical agencies to recognize the social economy as an important field to investigate longitudinally and in comparison with the "rest of the economy." In this book, with reference to said CIRIEC project, we gathered expert opinions and advice about key issues regarding the statistical methods and indicators for the social economy. In particular, we wanted to know: Why is it important to have statistics on the social economy? How are they produced? How might we better understand the social economy in the future? The goal was to arrive at an overview of the current state of knowledge on these topics. The objective was not so much to describe the technical aspects of methods or tools but rather to discuss their contributions and limitations, and to identify avenues for future research in the field of statistical production. Overall,

---

[1] More than 190 speakers from 25 European countries met in Rome on November 17-18, 2014 at the conference *Unlocking the Potential of the Social Economy for EU Growth*, organized under the auspices of the European Commission and the European Parliament. Institutional representatives, social economy practitioners, stakeholders, policy-makers, scholars and experts met to discuss the key issues affecting the development of the social economy in Europe. See: http://www.socialeconomyrome.it/.

[2] This project comes as a continuation of a previous CIRIEC working group on the methods and indicators for the evaluation of the social economy, which led to the publication in this collection of Peter Lang of *The Worth of the Social Economy. An International Perspective* (Bouchard, 2009).

this book intends to help researchers and decision-makers embrace this topic with a better understanding of what is at stake.

2. Context

In recent years, considerable work has been done to map the social economy, work that nevertheless yields a wide range of notions of this type of economy and a variety of methodologies for grasping its weight, size and scope. This reflects the diversity of the economic and political contexts in which the social economy exists. It also reveals the competing development models in which the social economy is called to play a role and the different paths of its institutionalization. As many countries have recently passed national legislations on the social economy, future advancement in this field can be expected.

Pioneer research associated with CIRIEC in the early 1990s exposed the presence and relevance in various national settings of the social economy concept and outlined its core identity (Defourny and Monzón, 1992). In 1997, Eurostat published the *Report on the Cooperative, Mutualist and Associative Sector in the European Union* (Eurostat, 1997) and, in 2006, CIRIEC was commissioned by the European Economic and Social Committee to produce a mapping of *The Social Economy in the European Union*, which was updated in 2011 (Monzón and Chaves, 2008 and 2012).[3] Another set of initiatives by the Johns Hopkins University on the nonprofit and voluntary sector in the late 1990s led in 2003 to the publication by the United Nations of the *Handbook on Nonprofit Institutions in the System of National Accounts* (Salamon, 2010). In order to build on and complement this handbook, the European Commission entrusted CIRIEC with the task of writing the *Manual on the Satellite Accounts of Cooperatives and Mutual Societies*. Published in 2006 (Barea and Monzón, 2006), the manual has so far been used by five European countries (Belgium, Bulgaria, the Republic of Macedonia, Serbia and Spain). Finally, recent research conducted for the European Commission produced a mapping of social enterprises in 29 European national contexts (Wilkinson, 2014).

All of these publications offer considerable knowledge about the sector and on how to measure it, with each addressing a particular aspect of the field. Some focus on the non-profit components (all voluntary non-profit organizations) and others on the market-based components (NPOs

---

[3] Study commissioned in 2006 and again in 2011 by the European Economic and Social Committee (EESC) and published in all the official languages of the European Union. The first study covered the 25 member states of the European Union. The update included two new member states, Bulgaria and Romania, and the two acceding and candidate countries, Croatia and Iceland.

with commercial activities, cooperatives, mutual societies). Some exclude certain core components of the traditional social economy and concentrate on the new social economy or the solidarity economy, or cover a very wide spectrum of legal statuses, including for-profit enterprises that pursue a primary social aim. Methodologies and indicators vary from one study to another and include satellite accounts, national observatories, impact assessments (social, economic, environmental) and longitudinal and demographic surveys. From an international perspective, the social economy remains a polysemic phenomenon spanning a wide range of practices and notions (Lévesque and Mendell, 2005; Chaves and Monzón, 2008 and 2012). For example, notions of social enterprise, social business and social entrepreneurship that began being in use as of the 1990s still carry a variety of meanings (Defourny and Nyssens, 2010; Wilkinson, 2014).

This makes it difficult to delineate and follow for statistical purposes the evolution of the whole of the social economy in a scalable fashion and in a way that allows for international comparability. Different tools and methodologies each have their own utility and limitations. Discussions about definitions might blur the legibility of the field, potentially undermining the social and political recognition of the social economy. In that context, it seems important to review the methods and their impact on the representations of the social economy, particularly in terms of outlining the field, nomenclature of activities and appropriate indicators. Lessons from these experiences have to be analyzed, especially in view of strengthening the capacity to develop coherent, rigorous and scalable information about the social economy as a whole and with respect to the nuances between the different notions or components of the sector.

3. Conceptual framework

This project adopts an open yet precise conceptual orientation. First, we have a broad definition of the social economy, allowing us to appreciate the scientific and empirical contributions brought forth by researches focusing on different components of the phenomenon. Second, while we recognize the value of standard economic indicators for measuring and comparing the social economy within national statistical accounting systems, we also acknowledge their limitations in assessing the full impact which the social economy has on the wellbeing of societies and communities. Third, we agree that these two issues – definition and methodology – are closely related, whereby a broad and open definition leads to a plurality of methodologies, and standardized methodologies to a clear definition. What is at stake is the trade-off between legibility and legitimacy, as social realities seem to be flattened or reduced when aggregated to satisfy the requirements of statistical treatment or, conversely, to leave room for misinterpretation or even ignorance when looked at as a series of unrelated phenomena. These issues justify the need for appropriate methodologies and indicators to quantify the social economy. They also explain the whys and wherefores of the debates about them.

3.1. Definition

Other than the technical aspects of measurement, quantifying a phenomenon involves having an accepted notion of what the subject is, valid indicators of how it can be recognized and distinguished from other empirical phenomena, and solid methodologies to compare it in different geographical, institutional and temporal settings.

The social economy is sufficiently distinct to constitute a "sector" or domain of the economy (Defourny and Monzón, 1992; Evers and Laville, 2004), recognizable as it is by its organizational characteristics, institutional rules and particular relationships with the state and the market. It refers to activities and organizations based on the primacy to people over capital. It borrows from yet also distinguishes itself from the capitalist economy and from the public economy by combining modes of creation and administration that are private (autonomy and economic risk) with those that are collective (associations of people), and its final aim is not centered on profit but on the primacy of the social mission over gain maximization (mutual or general interest). Two conditions generally explain the creation and development of social economy organizations: the necessity to respond to unfulfilled significant economic needs or social aspirations; and the fact of belonging to a social group that shares a collective identity or common destiny (Lévesque, 2006; Vienney, 1994; Defourny and Develtere, 1999). The general consensus about the social economy is reflected in the definitions adopted by international networks such as the European Social Economy Charter[4] or the European Parliament (European Parliament, 2009) and, at the national levels, by the recently adopted legislative frameworks on the social economy in Europe (Wallonie, France, Portugal, Spain, Greece), Latin America (Ecuador), North America (Québec) and those upcoming in various countries of the world, including in Africa and Asia. It is by and large agreed that the social economy includes cooperatives, mutual societies and associations, and increasingly also foundations. All share a number of principles that can be summarized as follows (Monzón and Chaves, 2012: 19):

---

[4]   http://www.socialeconomy.eu.org/spip.php?article263.

- The primacy of the individual and the social objective over capital
- Voluntary and open membership
- Democratic control by membership (with the exception of foundations, as these have no members)
- Consideration of the interests of both the members/users and that of the general interest
- Defense and application of the principle of solidarity and responsibility
- Autonomous management and independence from public authorities
- Use of most surpluses to pursue sustainable development objectives, services to members and the general interest.

However, the association with the term and its scientific concept, mostly used in Latin Europe, Latin America and parts of North America, is not homogenous across all countries, especially when it comes to outlining and measuring the sector. This was observed across European Union countries (Monzón and Chaves, 2012: 38; Wilkinson, 2014) as well as within one and the same country, where the term and concept of the social economy is often used interchangeably with other terms and similar concepts. Among these are: solidarity economy, popular economy, third sector, nonprofit sector, voluntary sector, civil society sector, etc. In some cases, the social economy is not defined as a sector but rather as "a mode of entrepreneurship and of economic development."[5] Over the last decade, the growing number of references to the notions of "social enterprise," "social entrepreneur" and "social business" has generated new questionings about the identity of the social economy and the foundations on which it is built.

This is also reflected in the approaches by which the social economy is defined. The social economy may be defined from a number of angles. Among these are its definite or indefinite (hybrid) juridical components (Desroche, 1983); its rules of functioning, which have both similarities

---

[5] One example of this is a law recently adopted in France – *Loi No. 2014-856 du 31 juillet 2014 relative à l'économie sociale et solidaire*, adopted in July 2014 – that uses a large and inclusive definition of the "social and solidarity economy," namely one that goes beyond the traditional legal forms (cooperatives, associations, mutual benefit societies, foundations) and includes activities of commercial societies which, by virtue of their by-laws, conform to the principles of the social and solidarity economy (as set out in the law), the quest for social utility (also defined in the law) and three management principles relating to profit and asset distribution limitations. Retrieved from http://www.legifrance.gouv.fr/affichTexte.do?cidTexte=JORFTEXT 000029313296&categorieLien=id.

and dissimilarities with public or other private economic entities (Vienney, 1980, 1994); its dynamics of reciprocity and solidarity within a plural economy (Eme and Laville, 1994; Evers and Laville, 2004); its non-profit and voluntary character (Hansmann, 1987); its social and entrepreneurial character (Dees, 1998; Borzaga and Defourny, 2004; Nyssens, 2006); and its innovative function (Lévesque, 2006; Mulgan, 2006; Caulier-Grice *et al.*, 2010). These approaches consist of more or less coherent and unified theories or paradigmatic fields (Nicchols, 2010). While some overlap with one another, others diverge in their essential orientation (Evers and Laville, 2004; Bouchard and Lévesque, 2015).

One assumption of this project is that the definition issue, notwithstanding its institutional or scientific importance, may distract attention from another important issue, that of the capacity to establish a comprehensive statistical overview of a reality that is becoming increasingly important for the future of our economies. Rather than proposing that one definition or theory is better than another, we argue that the social economy can be looked at through a variety of lenses, offering different field depths depending on which portion of the picture is of interest to the observer. We therefore use the term "social economy" in a broad sense to include associations (non-profit organizations or institutions), cooperatives and mutual benefit societies, as well as organizations that are recognized for their participation in the social economy, such as foundations, community economic development organizations, workers' associations (e.g., *sociedades laborales* in Spain, informal workers-owned enterprises in Brazil) and other mission-driven enterprises. This project adopts an approach that encompasses social economy organizations and activities in a broad perspective (market and non-market components). In so doing, it also enables the identification of subsets in accordance with the definitions of the social economy currently in use at a broad level in given national settings.

3.2. Statistical indicators and tools

The social economy has a rich history. Taking various shapes, it is measured in a variety of ways in different countries (Monzón and Chaves, 2012) and in different studies (Bouchard *et al.*, 2011). Measuring and evaluating the social economy is a complex process due to its very nature, which poses challenges.

A first challenge is that even though standard measurement techniques capture part of the social economy, another part of it remains difficult to grasp. The social economy contributes in very specific ways to the economy and society. However, since the social economy often participates in new or underdeveloped sectors of activity, its contribution may even fall "under the radar" of observation. In a challenge to standard classifications, it associates an economic activity with a social purpose and proposes innovative ways of doing things that correspond to values of equity, equality and social justice. Hence, applying standard statistical tools does not easily account for the identity and operating mode of this sector of the economy, which is multiform, permeable with other sectors (with emerging hybrid forms of organizations) and complex (combining social missions with economic activities). The weight of the social economy extends far beyond its contribution to job creation, GNP and economic added value. In addition to the many spillovers and externalities it generates, a large part of what it produces does not yet have quantifiable and agreed-upon measures.

Another issue is that the production of statistics for the social economy tends to relate to broader concerns about national accounting and the evolution and role thereof in policy planning. Historically, statistics have always cumulated the characteristics of a proving device, greatly based on mathematical formalism, and of a coordination device, implemented namely by administrative registers and national surveys (Chiapello and Desrosières, 2006). As the etymology of the word *statistiks* reveals (Desrosières, 2008), the statistical line of reasoning lies between science and government. The social legitimacy of statistics was established not so much from the formal methodologies for producing them but from their capacity to be considered as a prerequisite for decision-making in vaster sociopolitical projects. Yet, critiques have been expressed about the manipulative use of statistics by government (Data, 2009) or the inadequacy of economic indicators to actually reflect the performance of economies (Gadrey and Jany-Catrice, 2012).

Because producing statistics about the social economy has only become of interest in the past two to three decades, the stage at which we are at the moment is probably similar to that of other topics or phenomena, such as unemployment (Salais, Baverez and Reynaud, 1986), immigration (Fassmann *et al.*, 2009), racial origin (Fassin and Simon, 2008) or visible minorities (Beaud and Prévost, 2009), which appeared only recently and are recognized in some jurisdictions but not in others. One might also think of the statistical definition of "small and medium-size enterprises," which varies from one country to another and evolves from one period to the next. Statistical categories are indeed subject to debate and controversies. Added to this is the general warning against mistakes in economic policy that ensue from overestimating the explanatory power of numbers (La Documentation française, 2014; Ogien, 2013). Moreover, there is a growing resistance against the generalization of quantification as a mode of control of new public management (Desrosières, 2014: 34).

As any social phenomenon, the social economy is a social construct. Its quantification and measurement will result in the hardening of its definition and a decrease of any ambiguity about it, which in turn contributes to its legitimization. Yet, this same process will also reduce and flatten its complexity and richness, again prompting reflection about the appropriateness of definitions and methodologies. Thus, the production of statistics about any social phenomenon may become an issue, as it represents both an opportunity and a threat to legitimizing the field. This then calls for a solid social and reliable scientific construct of the social economy. That construct needs to be rigorous but also transparent and open to social actors' participation in order to reach a high level of social legitimacy. The readers of this book should bear in mind, as they go through the different chapters, that this circumstance, or challenge, is not exclusive to the social economy.

4. Outline of the book

This book is composed of two parts. The first part examines methodological concerns and theoretical issues that need to be addressed and clarified in order to produce sound statistics about the social economy. The chapters look into indicators, qualification criteria of entities, classification systems and the international standardization of methodologies. The second part of the book explores what can be learned from specific studies. The chapters examine particular methodologies (satellite account, impact assessment, national surveys) and issues (research collaborations, informal and hybrid organizations) in different national settings (Belgium, Brazil, Canada, France, Japan and the United Kingdom). These cases, while not exhaustive, offer a good overview of some of the most important methods that are used and the questions that ensue when producing statistics on the social economy.

4.1. Methodological concerns in producing statistics on the social economy

The first part of the book begins by taking stock of how the social economy is presently being measured. We explore what representations ensue from the varied methodologies and examine the questions of identifying, qualifying and classifying the entities. We discuss issues of international comparison and of the appropriate measurement of the social economy production.

In the very first chapter of the book (Monzón), we pay tribute to the important contribution of researchers from CIRIEC in the conceptual and statistical marking of the social economy, particularly in Europe. This pioneer work owes to a large extent to researchers coming from the Spanish section CIRIEC-España, namely José Barea, José Luis Monzón and Rafael Chaves. This chapter summarizes work that led to the production of invaluable tools for producing and analyzing statistics on the social economy in Spain and in Europe.

The next chapter (Artis, Bouchard and Rousselière) examines how the different indicators used in social economy statistics worldwide give way to specific notions or representations of the social economy. It begins by presenting the authors' point of view of statistical indicators, considered as social constructs. Based on a comparative analysis of research conducted on the social economy over five continents, the chapter then lists the main data sources used, describing both their strengths and limitations. It further identifies the representations of the social economy that are suggested, more or less explicitly, by these indicators. The main trends are identified with regard to the methodologies and depictions of the social economy and are then analyzed in light of theoretical frameworks of the social economy. The focus is on the methods and results that were the most innovative. Lastly, the results are integrated into the more analytical and theoretical debates within the field of statistical production.

Outlining the size of the social economy involves identifying the entities that take part in it and those that do not. Chapter 3 (Bouchard, Cruz Filho and St-Martin) looks into this question by analyzing some of the most important statistical studies on the social economy conducted by academic and institutional researchers, public institutions and statistical offices in various parts of the world. The resulting conceptual frameworks for producing statistics about the social economy usually establish which type of entities, legal statuses and activity sectors are excluded and identify a cluster of qualification criteria and statistical indicators of social economy organizations. Typologies of organizations can also be determined from other criteria, such as the goals and missions or the modes of financing them. A conceptual framework for qualifying social economy organizations also allows assessing peripheral developments or trends in this field and to anticipate its progress.

Social economy entities, being either market or non-market producers, must be included in the standard classifications, and they are – only not completely. Chapter 4 (Archambault) examines where in the national accounts these social economy entities are classified, and what the advantages and drawbacks of these classifications are. Challenges mainly have to do with the fact that standard classifications serve to analyze goods more precisely than they do services. With regard to the non-market production of collective or divisible services, they also reflect the government sector more than that of non-profit institutions. Statisticians are obliged to resort to standard classifications when undertaking cross-country comparisons of the social economy. Yet they have to innovate and adapt existing classifications when seeking to analyze other fields of action, such as reporting on the social economy's specificity, values, volunteer work or alternative way of governance.

The objective of producing a harmonized and agreed-upon reference was in part met with the publication in 2003 of the *Handbook on Nonprofit Institutions in the System of National Accounts*, which established an officially sanctioned procedure for capturing the work of non-profit organizations in national economic statistics, followed with publication of the *ILO Manual on the Measurement of Volunteer Work* in 2011, an internationally sanctioned tool for gathering official data on the amount, character and value of volunteering. These two manuals refer to investigations led by the Johns Hopkins Center for Civil Society Studies. In Chapter 5 (Salamon, Sokolowski and Haddock), the principle investigators of this group recall this experience and share what lessons can be learned for others seeking to extend this analysis into broader segments of the social economy.

Whether to generate management indicators (profitability ratios, structure ratios, etc.) or to create statistics on a macroeconomic level, conventional measurements often prove to be incapable of providing an accurate quantitative understanding of what a social economy enterprise produces. The authors of Chapter 6 (Mertens and Marée) first review how the production of the social economy enterprise is presently taken into account by the conventions of national accounting. They later introduce the notion of "broadened production" in order to take into account all the dimensions of what the social economy enterprise really produces. Finally, they conclude by showing how this "broadened production" cannot be the object of a unique monetary measurement. The authors instead advocate for recognition of the complexity of the production activity of social enterprises and formulate proposals that support the measurement of this production within another framework.

4.2. What can be learned from specific studies

The first four chapters of this second part of the book offer an overview of methodologies in use in Europe (France and Belgium), Latin America (Brazil) and Asia (Japan). Through these, we learn about issues concerning the production of a satellite account, the role of private social actors and of collaborative research in the actual production of statistics. In addition, we learn how institutional divides and the absence of an umbrella organization can impact the representation of the social economy at the national level. The subsequent two chapters discuss specific issues regarding, for one, the evaluation of economic impacts in the particular case of cooperatives in Canada and, secondly, the factoring in of the hybridity of both concept and methodology when mapping *social enterprise* in the United Kingdom.

One of the important methodologies for grasping the size and scope of an economic phenomenon is the satellite account. A satellite account structures the quantitative information from the national accounts relating to a particular area of study by offering a coherent system of statistical information that may be used for the purposes of macroeconomic analysis. The United Nations established in 2003 the first procedure for capturing the work of non-profit organizations in national economic statistics. Belgium was one of the first countries to implement this methodology as early as 2004. In order to complement this handbook, the European Commission entrusted CIRIEC with the task of writing a manual on the satellite accounts of cooperatives and mutual societies, which was published in 2006 (Barea and Monzón, 2006). Its purpose was to establish a rigorous conceptual demarcation of the social economy enterprises to be studied in the satellite accounts. Funded by the European Union, a project was launched in 2010 to set up satellite accounts for cooperatives and mutual societies in five countries, namely Belgium, Bulgaria, the Republic of Macedonia, Serbia and Spain. Chapter 7 (Fecher and Ben Sedrine-Lejeune) of this book summarizes the Belgian 2011 report.

France can be considered as one important birthplace of the concepts of the social economy, recognized by national actors and government as early as the 1980s. It also has a National Observatory on the Social and Solidarity Economy (ONESS). INSEE (National Institute of Statistics and Economics Studies) has been helpful in the creation of this observatory, although data on the sector are still produced by multiple private actors. Moreover, despite a growing consensus on the weight of the social and solidarity economy in the overall economy, even leading to the appointment of a minister and to the adoption of a legislation, and despite longitudinal monitoring and a large body of data, many questions remain regarding the scope, the categories and the indicators. Chapter 8 (Demoustier, *et al.*) first outlines the stepping-stones leading to the production of statistics on the social and solidarity economy in France. Then, the authors expose the different stakeholders that took part in the process and their viewpoints on the matter. They conclude in showing the importance of key social actors in improving the quality of the data produced, contributing in return to its better public and institutional recognition.

In Latin America, the notion most in use it that of the solidarity economy. It encompasses a multitude of social segments, agents and institutions, including cooperative banks, services and goods exchanges based on reciprocity, business networks and countless informal or formal associations of people seeking to develop economic activities, create jobs and engage in solidarity-based relations, be it among themselves or in society at large. In Brazil, there has been a major boom of the solidarity economy in recent decades. Chapter 9 (Gaiger) presents the *National Mapping of the Solidarity Economy* in Brazil, which was completed in 2013 after more than three years of work. The mapping was conducted through collaborative research carried out by solidarity economy actors with the support of research institutions and government. This extraordinary effort of collaborative research tested a unique methodology (snowball effect) and provided valuable empirical material to overcome the lack of knowledge about this sector.

The social economy is also a growing reality in Japan, where it plays a significant role in many sectors of activity. However, given the powerful corporate sector and the commanding public sector of that country, the social economy in Japan remains somewhat eclipsed. It also lacks the identity as a sector and cohesion among organizations involved, resulting in low recognition by the government, media or academia. This lack of public recognition then results in a lack of comprehensive statistics. Chapter 10 (Kurimoto) explains that the most important reason for this lack of visibility in Japan is the institutional divide. The social economy sector is compartmentalized based on laws, ministries, industrial policies and tax regulations, among other areas. It also suffers from the absence of umbrella organizations. In addition, there is no consensus on the notion of the social economy among researchers, and there is a competition between the North American notion of nonprofit organizations and the European notion of the social economy. The chapter gives a short description of the existing statistics and concludes with recommendations for producing comprehensive statistics of the social economy in the context of Japan.

Since 2008, there has been an increased interest in the impacts which social economy organizations, mainly cooperatives, have on the economy and on the communities in which they operate. Chapter 11 (Uzea and Duguid) presents a review of published literature on the economic impact of cooperatives in order to identify potential challenges and provide preliminary insights into how they can be addressed. These difficulties fall into three main categories, which correspond to the three stages of an impact analysis. They are (1) data collection issues, such as obtaining access to existing microdata, standardizing economic activity data across existing datasets, etc.; (2) data analysis issues, such as identifying the unit of analysis, adjusting the standard impact methods to suit the specifics of cooperatives, accounting for the unique outcomes of cooperatives, and examining the distribution of impacts, and (3) issues related to the interpretation of results, such as defining the counterfactual state and comparing the impact of cooperatives to that of the non-cooperative sector or across time. The conclusion of the chapter puts forward a number of ideas for future research.

While the concept of the *social enterprise* carries various meanings in different parts of the world, it is generally understood to have a more or less hybrid form of organization, such as a non-profit with commercial activities, a cooperative with primary social goals, or a for-profit enterprise with social values. Chapter 12 (Spear) examines the challenges, both conceptual and methodological, of mapping social enterprises given their hybrid form. These include difficulties in defining operational criteria, the matching to varied sampling frames, and making judgments about boundary cases. The chapter draws on other experiences of mapping non-profits and the social economy in order to obtain a greater understanding of mapping the social enterprise sector. Two concepts are introduced – hybridity and boundary cases – which are then used to discuss the mapping of the voluntary and non-profit sector and of social enterprises in the United Kingdom. The author concludes with comments on how certain measures could improve the operationalization of the definition of social enterprise prevailing in the United Kingdom as well as the mapping of the sector.

The conclusion of this book summarizes the key findings of its study and offers reflections addressed to policy designers, academic and institutional researchers and social economy actors.

References

Anheier, H. K., Knapp, M. R. J. and Salamon, L. M., *"No Numbers, no Policy – Can Eurostat Count the Nonprofit Sector?,"* in S. Saxon-Harrold and J. Kendall (eds.), *Researching the Voluntary Sector*, Tonbridge and London, Charities Aid Foundation, 1993, pp. 195-205.

Barea, J. and Monzón, J. L. (eds.), *Manual for Drawing up the Satellite Accounts of Companies in the Social Economy: Co-operatives and Mutual Societies*, Brussels, European Commission, D.G. for Enterprise and Industry and CIRIEC, 2006.

Beaud, J.-P. and Prévost, J.-G., *L'ancrage statistique des identités: les minorités visibles dans le recensement canadien*, Montréal, Université du Québec à Montréal, Centre de recherche sur les innovations, la science et la technologie (CIRST), No. 99-06, 2009.

Bouchard, M. J., *The Worth of the Social Economy, An International Perspective*, Brussels, Peter Lang, CIRIEC collection Social Economy and Public Economy, 2009.

Bouchard, M. J., Cruz Filho, P. and St-Denis, M., *Cadre conceptuel pour définir la population statistique de l'économie sociale au Québec*, Cahiers de la Chaire de Recherche du Canada en Économie Sociale, R-2011-01, Montréal, Canada Research Chair on the Social Economy / CRISES, 2011.

Bouchard, M. J. and Lévesque, B., "Les innovations sociales et l'économie sociale: nouveaux enjeux de transformation sociale," in J. Defourny and M. Nyssens, *Analyse socioéconomique du tiers secteur*, Brussels, De Boeck, forthcoming 2015.

Borzaga, C. and Defourny, J. (eds.), *The Emergence of Social Entreprise*, London, Routledge, 2004.

Caulier-Grice, J., Kahn, L., Mulgan, G., Pulford, L. and Vasconcelos, D., *Study on Social Innovation*, London, Young Foundation, Social Innovation eXchange (SIX) and Bureau of European Policy Advisors, 2010.

Chaves, R. and Monzón, J. L., "Beyond the Crisis: the Social Economy, Prop of a New Model of Sustainable Economic Development," *Service Business*, Vol. 6, No. 1, 2012.

Chiapello E. and Desrosières A., "La quantification de l'économie et la recherche en sciences sociales: paradoxes, contradictions et omissions. Le cas exemplaire de la positive accounting theory," in F. Eymard-Duvernay (ed.), *L'économie des conventions. Méthodes et résultats*, Paris, La Découverte, Coll. Recherches, Tome 1, 2006, pp. 297-310.

Data, L., *Le grand truquage. Comment le gouvernement manipule les statistiques*, Paris, La Découverte, Coll. Cahiers Libres, 2009.

Dees, J. G., "Enterprizing Nonprofits," *Harvard Business Review*, Vol. 76, No. 1, 1998, pp. 54-67.

Defourny, J. and Develtere, P., "The Social Economy: The Worldwide Making of a Third Sector," in J. Defourny, P. Develtere, B. Fonteneau, *et al.* (eds.), *The Worldwide Making of the Social Economy. Innovations and Changes*, Leuven, Acco, 2009, pp. 15-40.

Defourny, J. and Monzón, J. L. (eds.), *The Third Sector. Cooperative, Mutual and Nonprofit Organizations*, Brussels, De Boeck-Université/CIRIEC, 1992.

Defourny, J. and Nyssens, M., "Conceptions of Social Enterprise and Social Entrepreneurship in Europe and the United States: Convergences and Divergences," *Journal of Social Entrepreneurship*, Vol. 1, No. 1, 2010, pp. 32-53.

Desroche, H., *Pour un traité d'économie sociale*, Paris, Coopérative d'information et d'édition mutualiste, 1983.

Desrosières, A., *Gouverner par les nombres. L'Argument statistique II*, Paris, Les Presses de l'École des Mines, 2008.

——, *Prouver et gouverner. Une analyse politique des statistiques publiques*, Paris, La Découverte, 2014.

Documentation (La) française, *Le culte des chiffres*, Problèmes économiques, No. 3090, 2014.

Eme, B. and Laville, J.-L. (eds.), *Cohésion sociale et emploi*, Paris, Desclée de Brouwer, 1994.

Evers, A. and Laville, J.-L., "Defining the Third Sector in Europe," in A. Evers and J.-L. Laville, *The Third Sector in Europe*, Cheltenham, UK/Northampton, MA, USA, Edward Elgar, 2004, pp. 11-42.

European Parliament, *European Parliament Resolution of 19 February 2009 on Social Economy (2008/2250 (INI))*, 2009.

European Parliament and European Commission, *Social Economy Conference: Unlocking the Potential of the Social Economy for EU Growth*, Rome, November 17-18, 2014. http://www.socialeconomyrome.it/.

Eurostat, *Report on the Cooperative, Mutualist and Associative Sector in the European Union*, Luxembourg, European Commission, 1997.

Fassin, D. and Simon, P., "Un objet sans nom. L'introduction de la discrimination raciale dans la statistique française," *L'Homme*, No. 187-188, 2008, pp. 271-294.

Fassmann, H., Reeger, U. and Sievers, W. (eds.), *Statistics and Reality. Concepts and Measurement of Migration in Europe*, IMISCOE Reports, Amsterdam University Press, 2009.

Hansmann, H., "Economic Theories of Nonprofit Organizations," in W.W. Powell (ed.), *The Nonprofit Sector: A Research Handbook*, New Haven, CT, Yale University Press, 1987, pp. 27-42.

Lévesque, B., "Le potentiel d'innovation et de transformation de l'économie sociale: quelques éléments de problématique," *Économie et solidarités*, Vol. 37, No. 2, 2006, pp. 13-48.

Lévesque, B. and Mendell, M., "L'économie sociale, diversité des définitions et des constructions théoriques," *Interventions économiques*, No. 32, 2005. http://interventionseconomiques.revues.org/852.

Monzón, J. L. and Chaves, R., "The European Social Economy: Concept and Dimensions of the Third Sector," *Annals of Public and Cooperative Economics*, Vol. 79, No. 3-4, 2008, pp. 549-577.

——, *The Social Economy in the European Union*, Brussels, European Economic and Social Committee, 2012.

Mulgan, G., "The Process of Social Innovation," *Innovations: Technology, Governance, Globalization*, Vol. 1, No. 2, 2006, pp. 145-162.

Nicchols, A., "The Legitimacy of Social Entrepreneurship: Reflexive Isomorphism in a Pre-Paradigmatic Field," *Entrepreneurship Theory and Practice*, Vol. 34, No. 4, 2010, pp. 611-633.

Nyssens, M. (ed.), *Social Enterprise*, London, Routledge, 2006.

Ogien, A., *Désacraliser le chiffre dans l'évaluation du secteur public*, Versailles, Éditions Quae, 2013.

Rousselière, D. and Bouchard, M. J., "À propos de l'hétérogénéité des formes organisationnelles de l'économie sociale: isomorphisme vs écologie des organisations en économie sociale," *Revue canadienne de sociologie*, Vol. 48, No. 4, 2011, pp. 414-453.

Roy, M., "Working Group 7: The Contribution of Research, Education and Statistics," in European Commission and European Parliament, *Social Economy Conference: Unlocking the Potential of the Social Economy for EU Growth*, Rome, November 17-18, 2014. http://www.socialeconomyrome.it/.

Salais, R., Bavarez, N. and Reynaud B., *L'invention du chômage*, Paris, Presses Universitaires de France, 1986.

Salamon, L. M., "Putting the Civil Society Sector on the Economic Map of the World," *Annals of Public and Cooperative Economics*, Vol. 81, No. 2, 2010, pp. 167-210.

Stiglitz, J., "Moving Beyond Market Fundamentalism to a More Balanced Economy," *Annals of Public and Cooperative Economics*, Vol. 80, No. 3, 2009, pp. 345-360.

Stiglitz, J., Sen, A. and Fitoussi, J.-P., *Report by the Commission on the Measurement of Economic Performance and Social Progress*, France, Commission on the Measurement of Economic Performance and Social Progress, 2009.

Utting, P., van Dijk, N. and Matheï, M.-A., *Social and Solidarity Economy. Is There a New Economy in the Making?*, Geneva, United Nations Research Institute for Social Development, Occasional Paper No. 10, 2014.

Vienney, C., *Socio-économie des organisations coopératives. Formation et transformations des institutions du secteur coopératif français*, Paris, Coopérative d'information et d'édition mutualiste, 1980.

——, *L'économie sociale*, Paris, La Découverte, 1994.

Wilkinson, C., *A Map of Social Enterprises and their Eco-Systems in Europe. Executive Summary*, Report submitted by ICF Consulting Services, Brussels, European Union, 2014.

PART I

METHODOLOGICAL CONCERNS IN PRODUCING STATISTICS ON THE SOCIAL ECONOMY

The Pioneer Work of CIRIEC-España

José Luis MONZÓN

*Professor, Universitat de Valencia, and President,
Scientific Commission of CIRIEC-España*

Researchers from CIRIEC (Centre International de Recherches et d'Information sur l'Economie Publique, Sociale et Coopérative/ International Centre of Research and Information on the Public, Social and Cooperative Economy) have made important contributions to the social economy research field, namely to finding a conceptual and statistical definition of the social economy, particularly in Europe. This chapter offers a rapid overview of the pioneering work of researchers coming from the Spanish section of CIRIEC, CIRIEC-España. The first section recalls the stepping-stones of the creation of a Spanish observatory on the social economy. Section two and three present the methodology for drawing statistical portraits of the social economy, first applied to Spain and later on to other European countries. Section four explains how mapping the social economy in Europe benefited from this strong expertise as well as from the contribution of CIRIEC international network of researchers across Europe.

1. The creation of an observatory on the social economy[1]

Academics associated with CIRIEC-España look back on a long tradition of research on the social economy. They have contributed to the identification of the purpose of general interest of this economy; to the characterization of its distinctive features in terms of initiative (bottom-up), governance (democratic), ownership (collective) and surplus distribution (limited); and to the analysis of its permeable frontiers (Monzón, 1987; 1992; 1997; 2006). This conceptualization then served as a reference for the first CIRIEC international working group aiming to identify and compare the social economy in different national settings (Defourny

---

[1] Excerpts are taken from the websites of the Observatorio español de la economía social http://www.observatorioeconomiasocial.es/ and the Observatorio Iberoamericano del Empleo y la Economía Social y Cooperativa http://www.oibescoop.org/presentacion.php.

and Monzón, 1992). In the 1990s, the research of CIRIEC-España culminated in the production of the *White Paper on the Social Economy* (Barea and Monzón, 1992), and in 1995, in response to a mandate from the European Commission, in the production of a satellite account of the social economy in Spain (Barea and Monzón, 1995). In 2005, building on decades of research on the social economy in Spain, CIRIEC-España launched an observatory – the Observatorio español de la economía social – under the joint leadership of professors José Barea and José Luis Monzón and with the help of the Spanish Work Ministry (Ministerio de Empleo y Securidad Social), IUDESCOOP (Instituto Universitario de Economía Social y Cooperativa / University Institute of the Social and Cooperative Economy) at the University of Valencia, social economy groups and associations, as well as Red ENUIES, a network of Spanish university institutes and centers working on the social economy. The observatory has since become an indispensable instrument for the study and dissemination of data on the social economy in Spain. Its initial objectives were to assess the scope of the social economy, which it pursued by developing a rigorous theoretical framework for identifying the participating entities (Barea and Monzón, 1992). This methodology has permitted the observatory to quantify the social economy and evaluate its contribution to the Spanish economy from a longitudinal perspective (Monzón, 1995; 2010). The project, conducted with scientific rigor, is equally committed to society and to the social economy. Over the last few years, the Barea-Monzón research team realized a number of satellite accounts of specific parts of the social economy, such as foundations (Barea and Monzón, 2011a), third sector of social action (Barea and Monzón, 2011b), special work centers and the sector for people with disabilities (2008). José Barea passed away in September 2014, leaving an important heritage of methodologies in matters of national statistics on the social economy.

In 2008, OIBESCOOP (Observatorio Iberoamericano del Empleo y la Economia Social y Cooperativa/Iberoamerican Observatory of Employment and the Social and Cooperative Economy) was created under the leadership of IUDESCOOP and a consortium of institutions, among them CIRIEC-España, FUNDIBES (Fundación Iberoamericana de Economía Social/Iberoamerican Foundation of the Social Economy) and the University of Chile. Objectives of OIBESCOOP are to contribute to the sustainable economic development in Latin America by creating jobs and wealth through business initiatives of the social economy. Its aim is also to promote the development of a common ground, both theoretically and practically, for the social economy in Latin America, Spain and Portugal. The institution is under the responsibility of José Luis Monzón, who acts as CEO, while members of the board include José Maria

Perez de Uralde (CEO at FUNDIBES and the coordinator of activities in Latin America) and Mario Radrigán (University of Chile). The observatory relies on a network of social economy experts from each country. Moreover, OIBESCOOP has also established portraits of the social economy for a number of countries from Latin America and North Africa as well as for Spain and Portugal (Chaves and Perez de Uralde, 2013).

2. A methodology for drawing statistical portraits of the social economy

The CIRIEC-España scientific commission has developed a conceptual framework that facilitates the identification of the social economy as an institutional sector of a country's national accounting system, namely in concordance with the United Nations System of National Accounting (SNA) and the European System of National and Regional Accounts (ESA). The macroeconomic data aggregated by means of this framework proves to be essential for formulating public policies concerning the social economy. Applying both the ESA methodology and basic cooperative principles, the framework identifies two broad subsectors of the social economy: a) the market or entrepreneurial sector, which includes private enterprises with democratic governance (one person, one vote) and that have a benefit distribution scheme not linked to share capital ownership, and b) the non-market subsector, which includes private non-profit institutions serving households (NPISH).

This methodology proposes a new conception of the non-profit sector, namely one that is consistent with the definition of the social economy. The NPISH sector is comprised of non-profit organizations that provide goods or services to households (i.e., the general public) and that are funded predominantly by voluntary contributions from households, government grants and income from property. Thus, NPISH does not include the social economy non-profit institutions that were created, and subsequently financed, not only by non-financial associations or public administrations but also by financial corporations. Such institutions are presently classified in the institutional sector that corresponds to the entities that created them. Hence, the definition provided in the framework developed by CIRIEC-España is better able to capture the scope of the social economy than the ESA. In addition to incorporating NPISH as defined by the ESA, this definition includes: social services financed by other agents, foundations, other non-profit institutions financed by financial corporations and non-financial institutions, mutual insurance companies and non-market social economy groupings. Moreover, with this definition, the aggregation of all non-profit institutions into a new institutional sector is done with caution so as not to distort the main goal of the social economy, which is to serve people. For this reason, the framework does not count as entities of the social economy those non-profit institutions that are market producers engaged in producing goods and non-financial services, financial intermediation or auxiliary activities of financial intermediation, business associations, or those funded by voluntary parafiscal fees paid by non-financial corporations or financial institutions in exchange for the services provided by them.

Despite the common denominator "service to households" that is shared by this subsector of private non-profit institutions, this new composition of the institutional sector no doubt consists of a heterogeneous group of agents who are, moreover, of unequal importance from the standpoint of their social utility. However, based on what criteria are economic activities considered to be socially useful, or not? In general, activities are considered to be useful if they produce and distribute social goods that satisfy social needs not met by the public sector and the traditional market business sector. As such, these activities are performed by the so-called "third sector of social action," which consists of private non-profit institutions whose activity is focused on the production and distribution of the aforementioned social goods. Goods are considered to be "social," in turn, if there is broad social and political consensus that they are essential for a decent life and that they must therefore be accessible to all people irrespective of their income level or purchasing power. It follows that public authorities are then expected to provide for their production and distribution and ensure that they are either free or accessible well below market prices. The observatory of CIRIEC-España examines this broad range of enterprises and organizations comprised of the democratic and participatory enterprises dedicated to solving social problems through mutual solidarity as well as the non-market sector serving households. It should be noted that within this latter group, the so-called "third sector of social action" is increasingly developing links and joint action strategies with the business sector of the social economy.

3. Manual for drawing up the satellite accounts of companies in the social economy: cooperatives and mutual societies[2]

While a manual for drawing up satellite accounts of the non-profit sector did exist, there was none to embrace the whole of the social economy. In order to prolong and complement the United Nation's *Handbook on Non-Profit Institutions in the System of National Accounts*,

---

[2]   Excerpts are taken from Barea and Monzón, 2006.

the European Commission entrusted CIRIEC with the task of writing a *Manual for drawing up the satellite Accounts of Companies in the Social Economy: cooperatives and mutual societies*. Published in 2006, the manual was produced by José Barea and José Luis Monzón and two other CIRIEC researchers, Maite Barea and Hans Westlund (Barea and Monzón, 2006).

The manual first sets out criteria for identifying the entities of the social economy to be studied in the satellite accounts, and the classes of agents covered by those accounts. Those criteria are based less on legal and administrative aspects and more on the behavior of the social economy enterprises, including their resemblances and differences with each other as well as, as a group, with other economic agents. The manual identifies the two most significant agents in the social economy business sector, the cooperatives and mutual societies, on the basis of European Union documents and standards and on criteria established by organizations that represent the social economy in Europe. Subsequently, the manual establishes the criteria and requisites which business groups controlled by cooperatives and mutual societies, other similar companies in the social economy and certain non-profit institutions must fulfill, regardless of their legal form or designation, in order to be included in the satellite accounts. Finally, the various business agents in the social economy are located within the structure of the ESA 95 national accounts system. Five European countries have used this manual so far: Belgium,[3] Bulgaria, the Republic of Macedonia, Serbia and Spain.

4. Mapping the social economy in the European Union[4]

In 2006, the European Economic and Social Committee (EESC) commissioned CIRIEC to take stock of the social economy in the 25 member states of the European Union (Monzón and Chaves, 2008). That report was published in 2008 and was followed by an update in 2011, likewise commissioned to CIRIEC, that included the two new member states, Bulgaria and Romania, and the two acceding and candidate countries, Croatia and Iceland (Monzón and Chaves, 2012).

---

[3] Chapter of this book signed by Fabienne Fecher and Wafa Ben Sedrine-Lejeune summarizes the Belgian 2011 report.

[4] Excerpts are taken from the report *The Social Economy in the European Union* (Monzón and Chaves, 2012), from Social Economy Europe http://www.socialeconomy. eu.org/spip.php?article420 and from the European Economic and Social Committee document library https://dm.eesc.europa.eu/EESCDocumentSearch/Pages/redresults. aspx?k=social%20economy.

Both the initial report and the update were published under the direction of José Luis Monzón and Rafael Chaves and with support and advice from an expert committee composed of researchers from various countries. These are Danièle Demoustier (France), Roger Spear (United Kingdom), Alberto Zevi and Chiara Carini (Italy) and Magdalena Huncova (Czech Republic). The revised report was produced with the support of the CIRIEC national section Scientific Commission on Social and Cooperative Economy (presided at the time by Rafael Chaves) that made it possible to set up a very large network of correspondents and co-workers in all countries of the European Union and to benefit from CIRIEC's long record of research in key theoretical issues. One of the central objectives of the report, a comparative analysis of the current situation of the social economy by country, was achieved thanks to the decisive help of correspondents, including academics, sector experts and highly placed civil servants from the 27 member states and the two candidates for EU membership (Croatia and Iceland).

A further objective of the report was also to identify the core identity shared by all the organizations in the social economy sphere. Knowing what that core identity is would then allow to visualize and better understand the social economy. Questions addressed by the report include: Who are the social enterprises? How many are there? Where are they? How have they developed? How large or important are they? How do the public and governments see them? What problems do they solve and how do they contribute to the creation and equitable distribution of wealth and to social cohesion and welfare? Indeed, the report offers a very detailed and comprehensive overview of the social economy, including a definition of the social economy and an assessment of the extent of its activities. In addition, it highlights the quantitative and qualitative importance, in both economical and social terms, of the social economy sector in the European Union.

In total, some 15 researchers were engaged to analyze the situation in their respective countries and 63 European correspondents participated in the study as a whole, demonstrating the reach of the CIRIEC international network in Europe. It is also notable that many of these same correspondents had collaborated in a 2014 mapping of social enterprises in Europe.[5]

---

[5] Wilkinson, C. *A Map of Social Enterprises and their Eco-Systems in Europe. Executive Summary*, Report submitted by ICF Consulting Services, Brussels, European Union, 2014.

References

Barea, J. and Monzón, J. L., *Libro Blanco de la Economía Social en España*, Madrid, Ministerio de Trabajo y Seguridad Social, 1992.

——, *La Cuenta Satélite de la Economía Social en España: Una Primera Aproximación*, Valencia (Esp), CIRIEC-España, 1995.

——, *La Economía Social en España en el Año de 2000*, Madrid, CIRIEC-España, 2002.

—— (eds.), *Manual for Drawing up the Satellite Accounts of Companies in the Social Economy: Co-operatives and Mutual Societies*, Brussels, European Commission, D.G. for Enterprise and Industry and CIRIEC, 2006. http://ec.europa.eu/enterprise/policies/sme/documents/social-economy/.

Barea, J. and Monzón, J. L. (eds.), *Satellite Accounts for Foundations in Spain*, Madrid, Foundation of Savings Banks, 2011a.

Barea, J. and Monzón, J. L. (eds.), *Cooperative, Mutual Society and Mutual Provident Society Satellite Accounts in Spain*, Madrid, Spanish National Institute of Statistics for the European Commission, 2011b.

Barea, J. and Monzón, J. L. (eds.), *Economía Social e Inserción Laboral de las Personas con Discapacidad en el País Vasco*, Bilbao, CIRIEC-España, EHLABE Fundación BBVA, 2008.

Chaves, R. and Pérez de Uralde, J. M. (eds.), *La Economía Social y la Cooperación al Desarrollo. Una Perspectiva Internacional*, Valencia, Universitat de Valencia, 2013.

Monzón, J. L., *Las Grandes Cifras de la Economía Social en Espana*, CIRIEC-España, 2010.

Monzón, J. L. and Chaves, R., "The European Social Economy: Concept and Dimensions of the Third Sector," *Annals of Public and Cooperative Economics*, 2008, Vol. 79, No. 3-4, pp. 549-577.

——, *The Social Economy in the European Union*, Brussels, European Economic and Social Committee, 2008 and 2012.

Does the Social Economy Count? How Should We Measure It?

Representations of the Social Economy Through Statistical Indicators

Amélie ARTIS

Associate professor, Sciences Po Grenoble, France

Marie J. BOUCHARD

Full professor, Université du Québec à Montréal, Canada

Damien ROUSSELIÈRE

Full professor, AGROCAMPUS OUEST, France

Using methods known only to himself, our investigator
reported some very interesting statistics.
[Our translation]
Marcel Gotlib and René Goscinny, *Dingodossiers*[1]

Introduction

Statistics on the social economy are designed to quantify the relative weight of this type of economy (Salamon and Dewees, 2002), counteract its lack of visibility (Fecher and Sak, 2011) and improve overall knowledge and recognition of the field (UN, 2006). Quantification and

---

[1] French comic book series published in the 1960s.

evaluation are crucial steps for gaining a better understanding of the social economy and its place and role. Statistics also have a pragmatic function in that they allow to develop and assess public policy concerning the social economy at different territorial, national and international levels (Statistics New Zealand, 2004; London Economics, 2008).

However, given that the social economy is a relatively new concept in the field of national statistics, it faces many challenges. Among these are the identification of the statistical population (Bouchard *et al.*, 2011) as well as the use of appropriate classification categories (Bouchard, Ferraton *et al.*, 2008) and indicators. And, although standard economic indicators are able to accurately inform about some aspects of the social economy, such as sales figures or employment, they fail to shed light on aspects such as non-monetary production, the combination of market and non-market resources, the internalization of social costs and the reduction of environmental externalities. That said, the current production of social economy statistics does succeed in conveying certain aspects of the output of the social economy. This chapter examines how the different indicators used in social economy statistics worldwide give way to specific notions or representations of the social economy. We begin by presenting our point of view of statistical indicators, namely as social constructs. Then, based on a comparative analysis of research conducted on the social economy over five continents, we describe the main data sources used, showing both their strengths and limitations. Lastly, we identify the representations of the social economy that are suggested, more or less explicitly, by these indicators.

Building on that, we capture the main trends with regard to the identified methodologies and depictions of the social economy and analyze them in light of theoretical frameworks of the social economy. Here, our focus is on the methods and results that were the most innovative. Lastly, we integrate the results into the more analytical and theoretical debates within the field of statistical production.

Box 1: Methodology

Today, the production of social economy statistics encompasses a wide range of methods (satellite accounts, observations, surveys, portraits) and perimeters (non-profit sector, cooperatives and mutuals, social enterprises). How might we explain this diversity? How is the social economy taken into consideration in conventional economic statistics? Are these tools useful for quantifying the relative weight and evaluating the role of the social economy in the economy? To what extent does this production account for the specificities of the social economy? We attempt to respond to these questions starting from a study of statistical work produced worldwide on the social economy.

The countries covered by our literature review are England, Australia, Belgium, Brazil, Canada, Spain, France, New Zealand, Switzerland and Portugal. In addition, we take into account research on Europe conducted in the framework of international projects coordinated by the United Nations and the CIRIEC network. Due to language constraints,[2] only works in English, French, Spanish or Portuguese were studied.

For each research study, we identified the objectives of the statistical work, the range of observation, the indicators used, and the statistical and analytical results proposed.

The purpose of our study was to respond to the following main questions:

- To what extent do these statistical studies take into account the specificities of the social economy? How do they allow comparing the social economy with the rest of the economy? To what extent do the methodological choices influence the representations of the social economy?

- What are the explicit or implicit concepts that underpin these indicators? How do they reflect the particularities of the social economy? Are there any controversies surrounding them?

- What are the methods of data collection and processing? What is the quality of the indicators?

1. The production of social economy statistics: a question of method?[2]

For a few decades now, the social economy has been recognized as an increasingly important player in industrialized economies as well as in emerging or transition economies. In that context, actors on the ground as well as public decision-makers are experiencing an increasing need for solid knowledge of the empirical reality of the social economy, supported by statistical portraits of scope. Yet, given that the social economy is a relatively new object of study in statistics, it is still weakly codified. Furthermore, as the standard methodologies and indicators of national accounting systems generally fail to adequately account for this type of economy, other methodologies and indicators are needed.

One challenge in the production of social economy statistics is to quantify the economic weight of this economy in a way that is internationally comparable. Another challenge, in seeming contradiction to the latter, is to successfully convey the aspects of this type of economy that are not economic in the strict sense of the term as well as the role it plays in the different contexts where it takes root. This is important in

---

[2]   For example, studies published in German and Romanian could not be included.

order to prevent a simplistic reading of the social economy as nothing other than a government instrument or as a commonplace market agent (DiMaggio and Powell, 1983). The questions that then emerge concerning social economy statistics have to do with: 1) the quality of standard indicators compared to specific indicators; 2) the advantages of detailed approaches that are embedded in specific and local realities versus broader approaches that allow for comparison with the rest of the economy and with national social economies of other regions; 3) cost-benefit ratios of different methodologies on the basis of targeted objectives.

The primary function of statistical measures on the social economy is to measure the participation and relative weight of this type of economy within the overall economy, in other words, its share in the gross domestic product (GDP), in employment and in the different sectors of activity. By contrast, conventional statistical categories are designed to meet specific objectives and are therefore of limited value for measuring certain contributions of the social economy, in particular socio-economic and socio-political dimensions such as civic engagement, political activity and social innovation. At present, social economy statistics are mainly produced with two approaches: the top-down approach, implemented with satellite accounts, and the bottom-up approach, implemented with regional observatories or survey studies. Yet, their use is not without problems as they do not allow for sufficient comparability and, as we shall see, quantification results in differentiated representations of the social economy.

Overall, the evolution of social economy indicators is very similar to that of standard national accounting. Statistics of national accounts are coordination tools that are based on conventions (Chiapello and Desrosières, 2006; Gadrey, 2006). In all cases, statistical indicators are the result of institutional compromises, that is to say, of choices made in certain socio-economic, socio-political and socio-cultural contexts. According to Chiapello and Desrosières (2006), national accounting was initially developed in the 1940s and 1950s with the aim to establish tools to prove and measure reality. This approach was gradually replaced with the positive accounting theory, which assumes that reality, rather than pre-existing, is instead produced by quantification.

To that extent, statistical measurement is a social construct. Essentially, conventional national accounting indicators are a means to express the predominant notion of the economy in a fairly standardized way for a given period. Nevertheless, that notion of the economy is the subject of controversy, in particular concerning its approach to measuring economic performance and social progress (Stiglitz *et al.*, 2009). As such, indicators other than the GDP are today being proposed to measure the wealth and well-being of a country (Gadrey and Jany-Catrice, 2005). Despite a lack of international consensus as to their validity and usefulness, these tools attest to the necessity of going beyond standard economic measures, in particular for capturing other dimensions of well-being and progress, such as social well-being (autonomy, equity, health, social cohesion), leisure, the environment and happiness.

Many of these alternative indicators reflect a concern for the same issues as those addressed by social economy practices. Among these are the Human Development Index (HDI) of the United Nations Development Programme (UNDP), which measures health and education; the Genuine Progress Index (GPI) developed in the United States, which includes volunteerism; the World Values Survey based in Stockholm, Sweden, which concerns itself with values such as democracy; the Better Life Index of the OECD, which includes dimensions such as community and governance; and the implementation of sustainable development indicators following the Rio Summit of the United Nations (Gadrey and Jany-Catrice, 2005; Boarini, Yohansson and Mira D'Ercole, 2006; Bova, 2008; Gignac and Hurteau, 2011). This shows that the development and stabilization of new indicators is possible despite the wide range of notions of wealth. Thus, the social construction of economic indicators appears to increasingly incorporate what we might qualify as "the social." The evolution of standard indicators to measure the wealth of nations and the development of social economy indicators are two endeavors that may, one day, cross paths.

In the history of national accounting, observation has likewise been conceptualized with either the top-down logic, starting with aggregates, or the other bottom-up logic, starting from local observation. The different conceptual approaches with regard to the micro-macro links also play a key role (Vanoli, 2002). Top-down logic is based on aggregates and assumes that regularities observed at the macroeconomic level have a predictive value. Bottom-up logic, by contrast, is based primarily on the key economic agents and only secondarily on successive aggregations. These differences are also reflected in social economy statistics, where they manifest in the different approaches to developing indicators, ranging from satellite accounts to observation.

Recalling this is important, since the social economy, being new to the field of statistics, lacks knowledge of the subject matter. A first step is to examine what representations underpin the production of statistics on the social economy. A further step involves analyzing the ways of quantifying, classifying, collecting and processing data, all of which condition that which is measurable (Desrosières, 2008). Lastly, the statistical results must be analyzed in light of the specificities of the social economy as a type of economy that addresses political as well as social dimensions (Laville, Lévesque and Mendell, 2007).

2. Different methods of producing social economy statistics

Today, statistics on the social economy are produced on most continents (Africa, North America, Latin America, Asia, Europe, Oceania) and thereby include countries from the North (Belgium, Canada, Germany, Great Britain) and from the South (Argentina, Brazil, Pakistan, South Africa, Thailand). Those works focus on different subsets of the social economy, such as the non-profit sector; cooperatives, mutuals, associations and foundations; social enterprises (which may or may not be incorporated as classic for-profit businesses); and the solidarity-based economy (partly composed of informal businesses). Our review of the production of social economy statistics points to the use of different methodological approaches. These can generally be grouped into two types, the satellite account approach and statistical observation in the form of mappings. The studies also differ with regard to the subjects they deal with.

This diversity might be explained by the particular nature of the social economy, which defines itself by its rootedness in the needs and institutional dynamics of the setting in which it evolves. Moreover, at the national scale, these approaches are not necessarily opposed but rather complementary, as is the case in many countries (e.g., Canada, France and Italy). Finally, the observed variety is also a reflection of the debates on the methods, of the indicators and of national accounting.

2.1. The satellite account approach

Satellite accounts are statistical tables that are built to align with the framework of the general national accounts while integrating the specificities of a particular object. A satellite account "responds to a strong need to understand the economy of a field more precisely" [our translation] when that need cannot be satisfied within the general framework (Braibant, 1994). It thereby allows "to take into consideration the characteristics unique to a field under examination" [our translation] (Mertens, 2002: 247). This method allows complementing the general framework while preserving the coherence of the whole. It has been applied to a wide range of fields, such as research, transport, tourism, education, health, the social safety net, the environment and the social economy.

The first initiative in developing the approach was the construction of a methodology for the measurement and international comparison of the non-profit sector. Carried out by the Center for Civil Society at Johns Hopkins University, this undertaking, entitled "The Comparative Nonprofit Sector Project," was a research collaboration of international scope between different actors (research teams, statisticians, international institutions). The project contributed to the dissemination and adoption of a common framework of reference and culminated in the creation of the

Handbook on Non-Profit Institutions in the System of National Accounts, approved in 2002 by the United Nations (UN, 2006).[3] The application of this program of international comparison of satellite accounts from the non-profit sector rose from 13 countries in 1991 to 16 countries today.[4] More recently, the initiative has been complemented with the *Manual for Drawing up the Satellite Accounts of Companies in the Social Economy: Co-operatives and Mutual Societies* (Barea and Monzón, 2006), which has served to establish satellite accounts in many European countries (Belgium, Bulgaria, Spain, Macedonia, Serbia).[5]

The need for satellite accounts is explained by the increasing difficulty of measuring the role of the social economy in the national economy given the proliferation of agents from the social economy sector in the various institutional sectors[6] of national accounting. The data of a satellite account are compiled, following a specific methodology, from administrative data that are "by-products of some administrative function (such as tax collection)" (UN, 2006: 64). The first task is to create a statistical register of the entities (from the non-profit sector or the cooperative and mutual sector) starting from the identification files of businesses (such as the business master file maintained by the United States Internal Revenue Service or the SIRENE[7] database in France). Secondly, the data of the satellite account tables is derived from existing sources concerning revenue, output and paid salaries. In other words, they are comprised of existing aggregate data. Thirdly, the methodology suggests creating new data on the sector, either from specific administrative files or from the implementation of new data by way of surveys, such as the *Canada Survey on*

---

[3] See chapter by Salamon, Haddock and Sokolowski.

[4] Comparative Nonprofit Sector Project, http://ccss.jhu.edu/research-projects/comparative-nonprofit-sector/about-cnp (consulted on July 27, 2013). This project extended into a reflection on the measurement of volunteer work in the non-profit sector and was conducted in collaboration with the International Labor Organization. This research on the measurement of volunteer work intersects with other research on the improvement of data collection (Salamon and Dewees, 2002; Salamon, 2010).

[5] These studies were carried out following a call to tender issued by DG Enterprise and Industry of the European Commission: http://ec.europa.eu/enterprise/policies/sme/promoting-entrepreneurship/social-economy/#h2-5. See chapter by Fecher and Ben Sedrine-Lejeune in this book.

[6] National accounting is divided into the following five institutional sectors based on the similarity of their economic functions and the nature of their activities: non-financial companies, financial companies, public administrations, households and the rest of the world. See chapter by Archambault in this book.

[7] In France, all businesses are obliged to register with SIRENE (Système national d'identification et du répertoire des entreprises et de leurs établissements). The system identifies and indexes all businesses according to their principal activity, location, legal category and number of staff.

Giving, Volunteering and Participating (CSGVP) or surveys conducted among small structures (Barea and Monzón, 2006: 87). The last phase is that of the compilation of all collected data (UN, 2006).

The compilation of existing quality data and the creation of new data on satellite accounts contribute to the robustness and strengthen the stability of this method. The use of existing databases for other institutional sectors allows for consistency in the comparisons. In fact, the results integrate the evolutions of statistics, whereby changes are taken into consideration in an identical way for all institutional sectors. In this way, the method favors international comparability and the establishment of longitudinal follow-ups. Today, researchers use it for international and cyclical comparisons, such as the comparative analysis of the impacts of the 2008 crisis on the non-profit sector in France and in the United States (Archambault, 2010).

The satellite account method is based on an operational definition of the non-profit sector (Salamon, 2002) or the cooperative and mutual sector (Barea and Monzón, 2006) as well as on two specific accounting conventions, being the consideration of non-commercial output and volunteerism. The satellite accounts of the non-profit sector measure the contribution of this sector to the GDP by integrating market output, non-market output[8] and non-monetary output, the latter of which includes volunteerism. This type of study thereby reveals the diversity of resources used, which constitutes one of the specificities of the social economy, in particular the new social economy (Lévesque *et al.*, 2005) or the solidarity-based economy (Laville, 1994). However, the method does not allow an understanding of the benefits for individuals and society, in other words, the positive monetary and non-monetary externalities. Likewise, the price reference does not always allow measuring the global activity and its impacts (Mertens and Marée, 2012).

Two methods exist to measure volunteerism: the opportunity cost method and the replacement cost method (Prouteau and Wolff, 2004). The opportunity cost method consists of taking into consideration the remuneration that a volunteer could obtain if he/she dedicated his/her time to professional activities instead of the volunteer activity. This approach is categorized as a "loss of income" approach. The replacement cost method, by contrast, proceeds in terms of analyzing the "loss of spending" (Caillavet, 1998). The UN handbook (UN, 2006) recommends estimating the monetary value of volunteer time based on the average gross wage in

---

[8]   According to international conventions, non-market output is production provided free of charge or at prices that are economically insignificant (i.e., not covering the costs of production).

effect in the social services sector. It is also possible to refer to the average remuneration offered by associations in the field of activity, insofar as this information is available (Archambault, 1996; Mertens and Lefèbvre, 2004). Comparative research conducted at Johns Hopkins University shows that the monetary valuation of volunteer work is generally based on the average non-agricultural wage (Archambault, 2002) or the minimum legal wage (Prouteau and Wolff, 2004), if there is one.

As suggested by Prouteau and Wolff (2004: 47), the assigning of a monetary value to volunteer work is the subject of criticism, namely because it tends to frame volunteer work as free labor that complements a salary, thereby deflecting from the political and social dimension of volunteerism. Moreover, the assumption that volunteer work and paid labor are interchangeable does not apply to all contexts and masks the specificity of the associative form of production as a hybrid between salaried work and volunteer work. This issue is important because statistical research could provide clarification as to the role or impact of the social economy on the job market. For example, a study conducted in Canada shows that "[m]ost volunteers are engaged by organizations with relatively small staff sizes. There is little relationship between changes in staff sizes and changes in the size of an organization's volunteer complement" (Johnston *et al.*, 2004: 44).[9] It is thus important to question the notion that volunteer work and paid work are mutually interchangeable types of activities.

In the non-profit sector, there is strong support for the satellite account approach owing to its robust methodology, partnership-based construction, pragmatism and functionalism. As such, it is presently employed in sixteen countries. In the cooperatives and mutuals sector, the satellite account approach is more recent, although already applied in five European countries. The advantage of satellite accounts resides in their integration in a conventional framework that is shared by the economic actors, both private and public. The integration is facilitated by the use of aggregates that allow processing data at a lower cost, thereby forgoing the need to take account of the comprehensiveness of small organizations. The latter, although representing a significant population of the social economy, have only a weak influence on the overall volume. This characteristic allows for comparisons at the infra-regional level and encourages long-term studies. International comparison is also possible, as attested by research done at John Hopkins University.

---

[9] Another study shows that organizations that were a priori the most dynamic (defined as those having experienced growth over the past three years) are those that manage to mobilize different types of resources, including market (sales), non-market (grants) and non-monetary (volunteer) types (Rousselière and Bouchard, 2011).

The approach nevertheless remains fragile: "The use of satellite accounts in the framework of international comparisons will always be a risky and limited exercise given the very different demographics of the populations studied from one country to the next and the possible difficulties of capturing all businesses concerned" [our translation] (Fecher and Sak, 2011: 10). In addition, from a theoretical point of view, questions remain on the subject of certain assumptions underlying the exercise, such as monetary valuation, the predominance of an approach based on aggregates, or the relevance of existing systems of classification. According to some, transparency about the methodology employed paradoxically raises more questions than it resolves, as the clarification of all hypotheses and conventions used fuels criticism more easily (Lahire, 2005). As well, the monetization of social, cultural or environmental dimensions with fictitious sums of money that, although expressed in euros or dollars, do not correspond to real market (or financial) values, remains a challenge for the production of statistics.

2.2. Approach by specific observations and statistical surveys

Contrary to the satellite account approach, the approach by observations or surveys is not based on an international research project supported by international institutions. As a consequence, it is not promoted by means of established channels for international uptake and dissemination. Unlike global quantitative methods (such as national accounting), statistical observation seeks to understand internal specificities of organizations such as estimated time of work, women's share in the salaries of the social economy, and the distribution of salaries (e.g., in France, Switzerland, Brazil). Moreover, this approach focuses primarily on the local specificities of the social economy.

In France, observatories have been established since 2002 by representational bodies of social economy organizations, among them the Chambres régionales de l'économie sociale et solidaire (CRESS), the Conseil national des CRESS and the Institut national de la statistique et des études économiques (INSEE).[10] In an intermittent fashion, research centers also participate in these programs. The observatories develop research according to an infra-national dimension based on a decentralized structure and bottom-up aggregation. This approach favors taking into account the specificities of the social economy and a comparison with the other economic agents. The observatories highlight the relative share of the social economy within the local economy and in comparison with the private and the public sectors outside the social economy.

---

[10] See chapter by Demoustier, Braley, Guérin and Rault.

In contrast to the highly standardized satellite account approach, in this approach the differentiation and recognition of heterogeneity are promoted by means of a "customized" adaptation that takes regional realities into consideration. It is possible to define different boundaries of the social economy (e.g., the inclusion, or not, of social enterprises that do not have a cooperative, mutual or association legal status) and to develop specific indicators. Given this flexibility, this method is more frequently used for exploratory types of research. As such, it is also used to produce statistics on social enterprises, as has been done in the United Kingdom by Leahy and Villeneuve-Smith (2009). It moreover allows for the concrete observation of phenomena. For example, in the case of Switzerland, the observatory allowed studying the average salaries in the social economy relative to the national economy. Nevertheless, this method rests on the possibility, or lack thereof, to produce specific and usable data.

These observatories can be built on data derived from existing registers from statistical agencies, (such as the Centrale des Bilans (ConcertES, Belgium, 2008) or DADS (Déclarations annuelles des données sociales) (Bazin and Male, 2011)) or administrative agencies (such as the Business Register of the Canadian government) as well as from surveys – the whole with the idea of fostering a combination of registers and surveys. To complement the data produced by the national statistical agencies and to improve the taking into account of the specificities of the social economy, many studies use data collected by surveys conducted among the actors. In some cases, these surveys are of a substantial scope, such as a Brazil-based project (ANTEAG, 2009)[11] that observed 14,954 solidarity-based economy projects in 2005 or, more limited, a survey conducted in the United Kingdom comprising 962 telephone interviews with representatives of social enterprises (Leahy and Villeneuve-Smith, 2009). The methods are also very varied, ranging from telephone surveys to surveys by regular mail or email, with each positing *a priori* hypotheses on the framework of reference.

Unlike the satellite accounts that are based on a stable and homogeneous definition, the definitions of the scope of the observatories and surveys are more heterogeneous.[12] The legal statuses are often taken into account as a criterion of inclusion, although these are not necessarily the same or do not apply to the same situations from one country, or even region, to the next. Exclusions on the grounds of conceptual definitions of the social economy are also possible. Thereby, the social economy observatory in Switzerland excluded from its context certain cooperatives such as Migros or Coop, due to their high degree of institutionalization

---

[11] See chapter by Gaiger.

[12] See chapter by Bouchard, Cruz Filho and St-Denis.

(or trivialization of their cooperative identity). The research on the satellite accounts of social economy cooperatives and mutuals in Belgium examined only one portion of the cooperatives, namely those that are recognized as belonging to the social economy sector (Fecher, 2013).[13] It should also be noted that international comparisons are often difficult to perform due to the heterogeneity of the sources or units of observation. Moreover, many also argue that this trivialization, or dilution, of concepts comes at a price. Namely, a context in which any business can claim to be a social enterprise can cause a race to the bottom and dilute what is, in essence, the social economy (Social Economy Europe, 2014). The case of the UK is exemplary, in the negative sense, as the growth of the number of social enterprises was in fact due to the changing definition of the concept (Teasdale *et al.*, 2013). That said, advances have been made in the development, for the purposes of conducting statistical studies, of an operational definition of the social economy, as shown in the work of the European Commission on social entrepreneurship and "of the social economy at large."[14]

2.3. Differences and similarities of the methods

The two statistical approaches in the social economy, namely satellite accounts and statistical observations differ on many points: unit of observation; source of data; type of indicators; method; and objectives. Nevertheless, they face similar challenges, such as comparability or the development of pertinent indicators for the social economy.

In general, international comparisons are methodologically challenging (Lallement and Spurk, 2003). However, the use of the satellite accounts method offers advantages with regard to the homogeneity of data, comparability with the rest of the economy, and continuity over time. It is also dedicated to integrating the social economy into an international classification system that is recognized by everyone. The objective is to promote comparability through the use of recognized methods. In the great majority of cases, both methods use databases from national statistical agencies. However, they differ in that satellite accounts use aggregate data while statistical observatories use micro-data (i.e., the individual data of an entity). Thus, the latter involve specific surveys that are much more heterogeneous.

Moreover, debates persist on the development of indicators, the collection of data and the quality of sources. In fact, data that are available and

---

[13] See chapter by Fecher and Ben Sedrine-Lejeune.

[14] See the European Commission initiative, Social Entrepreneurship: http://ec.europa.eu/
internal_market/social_business/index_en.htm (consulted on August 30, 2013).

that meet the methodological demands are "painfully scarce" (Salamon, *et al.*, 2011). Furthermore, the use of data that was originally gathered by national statistical agencies implies a lower degree of control and less room to maneuver. Essentially, since specific data on the social economy do not exist, it is necessary to extract the information from existing files and to create a new statistical population. Yet, that process involves finding a conceptual and operational definition of the boundaries of that population. Likewise, the creation of new databases (e.g., to identify more closely the voluntary contribution of households (UN, 2006: 65)) comes up against a cost problem.

In addition, the approaches differ in their manner of taking into account small organizations. The satellite account approach of the non-profit sector focuses on overall volume. It proceeds by sampling the smallest organizations (often non-employing), tolerating a certain imprecision but seeking to be exhaustive with regard to the largest ones. By contrast, the observational approach has primarily the objective to create typologies of social economy organizations with the company, or establishment, as unit of observation. This allows better taking into consideration the emerging dynamics, in turn allowing the study of phenomena of small scales.

Further, satellite accounts as well as statistical observatories resort to the existing classification systems of a sector, which can pose difficulties in interpretation.[15] First, social economy organizations often engage in complementary activities, yet statistical processing focuses on the principal activity as indicated. In addition, the categories do not always correspond to the activities of social economy organizations, for example by remaining associated to public administrations. Moreover, many activities represented by the social economy form sub-groups that are too small to represent a statistical category of their own. Together, these form a residual category that is difficult to interpret. Consequently, numerous categories are empty and others are inappropriate. Lastly, by convention, the principal activity is the one that uses the most resources, which may thus change once non-market output and volunteer work are integrated. For example, in the case study on Montréal in the cultural domain, reclassifications were made between recreational or production activities (Bouchard, Rousselière *et al.*, 2008).

However, despite these differences, we observe certain coherence in the results and the indicators used. As such, the social economy is an economic actor in its own right that contributes to production and to employment. It is primarily present in activity sectors where it has a significant impact, such as, in some countries, agriculture, financial services and people

---

[15] See chapter by Archambault.

services. As we shall see in the following section, different conceptions of the social economy are underpinned by the methodologies employed.

3. The different conceptions of the social economy in the production of statistics

Our analysis of the production of social economy statistics identified three commonly-held conceptions: the social economy as an economic agent, the social economy as an economic model and the social economy as a territorial actor. We then put these three notions into perspective using standard economic reasoning. In a further step, we examined and described them in more detail, to different degrees and with regard to country-specific topics, in order to demonstrate the particularity of the social economy as a business form.

3.1. The selection of variables

The variables that serve to represent the social economy in statistics are mainly variables of an economic nature. Generally, the principal statistical works that were analyzed[16] mobilize:

- standard monetary variables coming from accounting and financial frameworks;
- supplemental monetary variables specific to the sector (non-market output, hours of volunteer work); and
- social and economically quantitative variables specific to the sector (number of members, number of volunteers, profile of members, etc.).

The variables correspond to traditional economic categories:

- economic and social performance: contribution to GDP, sales figure, gross added value, revenues/expenditures, number of entities, number of direct jobs, number of volunteers, total workforce, market share;
- internal organizational structure: employees (women, disabled, temporary, part-time/fulltime), volunteers, number of members;
- sector of activity or mission;
- relationship to the territory: implementation, regional distribution.

These different variables often overlap, as is the case with the distribution of equivalent full-time jobs by sector of activity or the distribution of structures by sector of activity.

---

[16] See the methodology presented in the introduction of this chapter.

The indicators are developed through an identification of the concept observed, followed by the breakdown thereof in many dimensions in order to isolate the individual variables (Lazarsfeld, 1965). In the case of the social economy, as in other fields (education, tourism), two processes complement each other: the use of standard indicators, fostering the integration with the other economic agents; and the development of specific indicators, allowing completion of the particular dimensions of the observed phenomenon in a differentiating manner. In the case of the social economy, these dimensions are, for example, the members and the volunteers. The standard indicators view the social economy in the same light as the other economic agents, in other words, from an economic perspective. The variables of wage earner and volunteer express the values of commitment or solidarity that characterize the sector. Improvements in that regard will require conceptual and methodological advances, for example as proposed with the notion of expanded production,[17] which integrates the entirety of realizations, results, and direct and indirect impacts.

Based on our review, impact indicators are rarely used in research. However, these questions are an object of interest on the part of the organizations themselves. For example, one study on the socio-economic impact of work integration social economy enterprises in Québec uses the inter-sector model of the national statistics agency to determine the direct, indirect and induced impacts of these businesses (Comeau, 2011). This field is still in an exploratory stage, with methods, although conceivable, not yet applied to the social economy.[18] Moreover, the question of the structuring effects on the economy remains open.

3.2. The social economy: many frameworks brought to light

Our analysis of the production of social economy statistics brings to light the presence of several notions of the social economy. Of those, the notion of the social economy as an economic actor prevails.

3.2.1. The social economy, a fully-fledged economic agent

In all the research examined, the social economy is primarily portrayed as an economic actor of the same order as the other forms of businesses (such as those from the private sector or the institutional sector of non-financial businesses). For example, the studies present the social economy primarily in terms of its GDP contribution, number of businesses, and contribution to added-value production: "The nonprofit sector accounts for an average of 4.5% of the GDP in the covered countries,

---

[17] See chapter by Mertens and Marée.
[18] See chapter by Uzea and Duguid.

roughly equivalent to GDP contribution of the construction industry in these countries" (Salamon *et al.*, 2013: 5).

Social economy research then applied this first aspect, which shows that the social economy is an economic actor like the others, to other concepts, such as non-market production. In satellite accounts, this has allowed to highlight a contribution of the social economy that often remains invisible, as it is not used by commercial businesses. Yet, the use of such concepts is fundamental for the social economy in that it "adds to the SNA[19] basis an estimate of the non-market output of 'market' NPIs [Not Profit Institutions] in the SNA corporations sectors. This adjustment is necessary in view of the fact that market NPIs, unlike other market producers, typically also have substantial non-market output that is not captured in their market receipts" (Tice *et al.*, 2002).

The study of the social economy as an economic actor reveals its contribution to the creation of wealth and identifies its part in the various sectors of activity. The different works showed the importance of the social economy in sectors such as people services (Tice *et al.*, 2002) or local services such as those related to sports and recreation (Johnston *et al.*, 2004). This sectorial distribution confirms the recognition of the social economy as an important, and sometimes even predominant, actor in relational, creative and financial activities: "The vast majority (nearly 75%) of nonprofit gross value added (GVA) is generated through service activities as opposed to expressive activities" (Salamon *et al.*, 2013: 7). Many analyses explain this by the fact that these services are co-produced in the interaction between producer and user, involving face-to-face, interpersonal social dimensions and an emphasis on proximity (Autès, 2006; Demoustier and Ramisse, 2000). In the social economy, "service relationships" in the sense of Gadrey (2003) take shape through the interaction of users in the co-production of the activity, thereby creating social proximity and trust.

In terms of inputs, the studies show that the social economy provides a substantial portion of jobs in the active population in the different economies. For example, from a macroeconomic perspective, the European social economy (associations, cooperatives, mutuals and foundations) comprises a significant portion of human and economic activity, employing more than 11 million people and representing 6.7% of the salaried population of the European Union (Chaves and Monzón, 2008: 21). In the non-profit sector, this dimension is also essential: "The nonprofit

---

[19] The System of National Accounts (SNA) is a comprehensive, consistent and flexible set of macro-economic accounts intended to meet the needs of government and private-sector analysts, policy-makers and decision-makers (UN, 1993).

workforce, including paid and volunteer workers, makes up 7.4% of the total workforce on average in the 13 countries on which full data are available. This places it ahead of a number of major industries, such as transportation and finance" (Salamon *et al.*, 2013: 4).

The cross-tabulation between the labor pool and sectors of activity demonstrate that the social economy has the capacity to be a main employer in sectors such as social services. The study of the social economy as employer also seeks to show the quality of that employment, mainly on the basis of the proportion of full-time jobs versus part-time jobs or its distribution by gender and socio-professional categories. These observations are guided by a range of hypotheses on the social economy as employer, including its capacity to create, maintain and sustain employment and its overall role in the labor market.

Moreover, research conducted over several years allows showing the employment dynamics in the social economy. For example, the analysis of associations in Belgium between 1998 and 2002 reveals that "all branches showed a non-negligible growth of employment in associations over the period examined, with the strongest growth noted for the area of culture (+45.4%) and social action (+25.8%) and the weakest in education (+2.6%)" [our translation] (Marée *et al.*, 2005: 57). Although the data from the different periods are comparable, methodological difficulties do exist with regard to the homogeneity of data.

As such, a study conducted on the distribution of social economy salaries in Switzerland showed that average salaries are largely above the minimum salaries demanded by the unions (APRÈS-GE, 2010). This study also analyzed the impact of public financing on job growth in each of the social economy sectors, being associations, cooperatives, mutuals and foundations. The results show that an additional Swiss franc is correlated to a higher job increase in associations (APRÈS-GE, 2010: 27) and that "financial grants significantly increase the number of employees in associations, and more so there than in cooperatives and foundations" [our translation] (APRÈS-GE, 2010: 28). These findings should probably be put into perspective given the strong endogeneity of the variables. Nevertheless, in another study that took this question into consideration, organizations' success in obtaining grants was shown to increase with the number of employees (Rousselière and Bouchard, 2011). This suggests that employment in the social economy in terms of quantity and quality is more complex than it appears, in particular due to the increasingly hybrid forms of employment and work (wage earner/volunteer).

Studies of the social economy referring to standard approaches utilize number of employees as an indicator, but they often neglect many other aspects of that economy, such as the economic contribution of solidarity or collective action. In the satellite account method of the non-profit sector, volunteer work is integrated as a specific form of work in the social economy. For this, volunteering is quantified in terms of hours and then compared to the hours worked by wage earners in the active population; thereafter it is valued and monetized on the basis of the principle of replacement cost (UN, 2006; Prouteau and Wolff, 2004). Volunteer work is thus measured as a free resource. This approach has spurred debate about the data sources and the possible ways to improve them (Salamon and Dewees, 2002), the quantification method (Salamon, 2010) and the means of evaluation (Prouteau and Wolff, 2004). That said, this approach remains important for highlighting the specificity of the social economy as an employer and for questioning the boundaries between activity, labor and employment, and between civic engagement and work. On the other hand, viewing volunteer work as nothing other than a resource undermines its role as creator of social ties within territories and as an indicator of civic commitment undertaken by individuals (Demoustier, 2002).

Some studies also adopted a more exploratory approach to examining other specificities. In the case of Brazil, for example, one study has researched collective entrepreneurship through the analysis of activities realized in a collective fashion, the frequency of meetings, and the democratic participation in organizations of the solidarity-based economy. The study attempts to quantify the forms of cooperation between the organizations of the solidarity-based economy, revealing the inter-cooperation brought to light by Vienney (1994) as a form of development that is particular to the social economy.

In these ways, the social economy is qualified as an economic actor on the basis of it being a creator of wealth in the production of goods and services according to a results-based (output) logic model; and as an economic actor that organizes labor and capital according to an input-based logic model.

3.2.2. The social economy, a unique economic model

The different statistical analyses of the social economy model focus primarily on its costs and revenues, with consideration of its specificities (e.g., no-cost items such as volunteer work, particular revenues such as contributions), and secondarily on its unique mix of private and public resources that underlines its similarity to public financing.

The satellite account approach proposes an evaluation of revenues coming from the sale of both market and non-market output, for example: In the U.S., "[a]s of 2007 the revenue of public benefit, non-profit organizations stood at slightly over $1.7 trillion" (Salamon, 2012: 8). Or: "NPIs received $76,639 million worth of income in 2006-07. The main source of income for market NPIs was sales of services ($19,591 million), whilst the main source of income for non market NPIs was volume based government funding ($4,253 million)" (Australian Bureau of Statistics, 2007: 5). Thereby, it is possible to measure the share of private market financing of the social economy so as to arrive at an estimate of its degree of financial autonomy and integration in the market production. This element is complemented by a determination of organizations' own resources, as in the studies on Brazil, where 65% of organizations' own resources are generated by members (ANTEAG, 2009). In the case of France, a survey realized by the Centre d'Économie de la Sorbonne shows that 12% of associations' resources are derived from contributions (cited in CNCRESS, 2012: 95).

This line of questioning is crucial for the social economy because it allows to confirm or rule out the idea that the sector is essentially dependent on grants, and to reaffirm the integration of the social economy in the economic sphere. For Switzerland, the above-mentioned study conducted by APRÈS-GE shows that the "social economy is not an economy of subsidized organizations […]. One third of APRÈS-GE member organizations are completely self-financed. The others receive public and private financial support for activities that serve the public interest" [our translation] (APRÈS-GE, 2010: 7). In a similar vein, the study on social enterprises in the UK examines their break-even points to identify those that are profitable and to analyze their performance (Leahy and Villeneuve-Smith, 2009). Some works also try to quantify the portion of revenue that is reinvested in an organization, as was done in the studies on Brazil and the UK, in order to express the importance of social action within entrepreneurial dynamics, to testify to the non-profitability and the non-appropriation of funds by individuals, and to thereby validate its contribution to the community.

With reference to the specificity of the social economy as a hybrid economic model (Laville, 1994), the issue consists in showing how this economy contributes to and impacts the GDP and the general interest through its different forms of private and public financing. In many cases (as in Canada or France), the statistical agency does not describe and itemize the different resources coming from a public source, listing them rather as grants. Yet, in the case of calls for tender or contractual agreements (public service delegation), the government approves resources in exchange for services rendered (unit cost of delivery), whereas subsidies are granted to support the existence of the organization or to help it bear its production or capital costs. However, the capturing and quantification of this phenomenon is complex because the interviewees tend to have wide-ranging responses (Bouchard, Rousselière *et al.*, 2008).

Thereby, the taking into account of revenues coming from the public sector implies identifying the contracts and contributions belonging to a market (or quasi-market) as well as the subsidies generated in the non-market rationale of redistribution. Some studies differentiate between the types of resources coming from public sources. For example, the study on the non-profit sector in Canada states that:

> We have grouped these into three categories: *fees*, which includes earned income from private payments for services, membership dues, service charges, and investment income; *philanthropy*, which includes individual giving, foundation giving, and corporate giving; and *government or public sector support*, which includes grants, contracts, and reimbursements for services to eligible third parties from all levels of government (Hall *et al.*, 2005: 15).

However, the resources are not explicitly distinguished by type in the remainder of the analysis.

In research based on the satellite account approach, it is possible to quantify the relative share of each type of financing. For example, with regard to the non-profit sector in Australia, it is estimated that "[s]ales of goods and services is the main source of income (61%) for all non-profit institutions. Income from transfers (33%) is also significant. Investment income is relatively small" (Australian Bureau of Statistics, 2004). Yet here as well, no distinction is made between quasi-markets and other forms of public money transfers.

The composition of resources often overlaps with the sectors of activity. The Australian case also brings to light the public/private financing model for advocacy and outreach organizations and associations working with aboriginal populations. The nature of the resources is also correlated with the size of the organization, as was done in the study on France, where the largest associations obtain the largest amounts of public financing because "public aid is very concentrated: 2% of associations receive 55% of public financing" [our translation] (CNCRESS, 2012: 94). Finally, further research points to the presence of reinforcing effects: the greater the share of public grants, the greater the resources coming from sales (Rousselière and Bouchard, 2011).

Satellite accounts also integrate charitable donations, itemized by donor and recipient (UN, 2006: 70). In the observation made in France on employment, this integration showed that 5% of association resources come from gifts (CNCRESS, 2012). In the case of Canada, gifts are also analyzed as indicators of civic engagement (Johnston *et al.*, 2004). For example, of all Canadian provinces, Québec was shown to have the lowest average amount of donations and the lowest rate of volunteerism (Hall *et al.*, 2009), yet its volunteer and non-profit organizations draw 60% of their revenue from the governments (as opposed to organizations in Alberta, which draw only 33% of their revenue from governments) (Johnston *et al.*, 2004).

Therefore, statistical representations of the social economy integrate the specificities of the social economy model in showing the diversity and the interlocking of the various available financial resources, namely private market resources (sales), public market resources (contracts), non-market private resources (gifts) and public resources (grants). These results should be considered in the debate on the autonomy and dependence of the social economy with regard to the public actor. In examining the nature and the portion of public financing in their total budget, Hall *et al.* (2005) establish three different positions of social economy organizations in relation to the state: partnership, autonomy and dependency. This theoretical framework seems pertinent for interpreting configurations on a per-country basis.

3.2.3. The social economy, a territorial actor

The different statistical studies propose a geographic distribution of the social economy through the quantification and localization of social economy activities taking place on the territories (in terms of number of businesses, volume of jobs and sector of activity). For example, the Canadian and French studies, based on local observation, developed an infra-regional approach for observing larger regions and a more ad hoc approach for observing smaller territories.

Three assumptions underpin this territorialized representation. One, the social economy is anchored in the territory, both through the activities it produces and distributes and through the engagement of its members in the governance of the organization. This territorial anchoring is often seen as a bulwark against the relocation of social economy organizations, strengthening their role as producer and local employer. Nevertheless, this perception is questionable, as locational choices in the social economy are also determined by economic strategies.[20] The second assumption is that with a local development approach, the social economy is the producer of amenities constituting a territorial resource, as it contrib-

---

[20] The inalienable character of the type of ownership is also a further fundamental characteristic of the social economy that reduces the risks of relocation. At the international level, relocations are rare. By contrast, closures of organizations on infra-national territories may occur in many fields of activities of the social economy, as was the case in France during efforts to consolidate the various departmental mutuals, or in Canada with the merger of small agricultural cooperatives.

utes to the attractiveness of the territory, be it in the productive sphere (productive activities and jobs) or in the in-place sphere with regard to people services (services to the population, recreation, sports or culture). The third assumption underpinning the above-mentioned territorial representation is that the social economy is more beneficial for people from modest or low-income social classes rather than the wealthier segments of the population.

The territorial approach shows that the social economy is anchored in particular territories. Indeed, it is estimated that the social economy is more present in rural territories or in those with poor economic indicators or several deprivations:

> Social enterprises are likely to be situated in areas of high multiple deprivation; 29 per cent are located in the 20 per cent most deprived wards and a further 20 per cent in the 20 to 40 per cent most deprived wards. However, social enterprise activity is not restricted to areas of deprivation and half of those (49%) identified operate in areas that would not be considered deprived (IFF Research, 2005: 3).

Although the correlation between a population's level of income and the location of the social economy has not been proven, it appears that the social economy is intrinsically linked to specificities concerning "territorial matrixes"[21] within the meaning of Itçaina *et al.* (2007). Many differences exist between urban and rural territories, as demonstrated in the study on Canada with regard to the financing of charities, the structure (asset to liability ratio) of which differs based on location: "The ratio of assets to liabilities, however, appears to be more favorable for rural charities: the average assets of rural charities are 2.5 times their liability, while the average assets of urban charities are only 1.9 times that of their liability" (Johnston *et al.*, 2004: 25).

However, the statistical analysis of the social economy by geographic distribution does not shed light on the causal relationships between social economy organizations and the territory. In that sense, interpretations of causality must be made with caution and should include complementary qualitative approaches. Long-term longitudinal approaches are also of interest in allowing to formulate hypotheses about these territorial links.

---

[21] The territorial matrix is a combination of cultural, historical, political and economic factors that allows to understand, explain and interpret the rootedness of the social economy in the territories and its contribution, or not, to the wealth and transformation of the territory (Itçaina *et al.*, 2007).

Conclusion

Today, statistical studies present the social economy as an economic actor in its own right and as a contributor to the economy of the countries studied.[22] This integration is not without methodological challenges. First, how might this economic form that is scattered across many sectors of activity and institutional sectors be taken into account? Secondly, how can these specificities be reflected within the framework of normalized accounts? A review of research published to date in different parts of the world permits us to assess the state of affairs of these questions from two angles. One concerns the theoretical debates on the social economy and on statistics itself, and the other concerns the contribution of the social economy within the economy as a whole, which in turn also points to shady areas.

Today, two approaches participate in the production of social economy statistics. The first, the satellite account approach, integrates the non-profit or the cooperatives and mutuals sectors into the national accounting books, in addition to capturing the non-market production and volunteer work. The second, the approach by observation and statistical surveys, measures the contribution of social economy organizations to a local or national economy while also underlining its specificities. The two approaches have the common objective to quantify and measure the contribution of the social economy to the economy at large. They also face similar challenges concerning international comparability and the collection and quality of pertinent data in the field. However, they differ on several points, namely the methods, boundaries and objects.

The satellite account approach subscribes to a top-down logic starting from data aggregated from national accounts, while the observatory approach favors a bottom-up logic starting from surveys and observations of specific data. Yet, as national accounts are adjusted with each other at an international scale, the satellite account approach facilitates comparability more quickly. However, this approach requires thorough knowledge of national statistics accounts. On the other hand, the observation founded on national or local definitions of the social economy allows sounding out the particularities of the output of the social economy and fosters a differentiated picture of national or local models. This reflects the fact that the social economy assumes a place and plays a role depending, among others, on the place and role of the state, the market and civil society. The question of

---

[22] The studies reviewed in this chapter include very few developing countries. In that sense, a whole other study would have to be undertaken on the difficulty of grasping the scope and the role of the social economy in such institutional contexts.

the boundaries of the social economy, in each society, raises historical and conceptual questions (Laville, Lévesque and Mendell, 2007).

As any new field, statistical production on the social economy raises many methodological questions on the collection of data and the development of indicators. The creation of new data with new indicators is a long and costly process. It also calls on the different actors (statisticians, researchers, social economy organizations) to reach a compromise concerning the scope of the study, the indicators and appropriate measuring tools. By contrast, when using existing data, the social economy is represented with categories that were initially developed for the rest of the economy. Thus, selecting a procedure for data collection involves weighing its advantages and disadvantages: the exploitation of existing data or the creation of new surveys; exhaustive data on a limited number of indicators or targeted data on many or complex indicators; data already aggregated but less adapted to the social economy, or data that is specific but sometimes too limited in number to allow for satisfactory levels of interpretation (e.g., in new sectors of activity) or simply containing the risk of a low response rate.

The statistical production on the social economy shows the contribution of this sector at three principal levels. One, the social economy is a fully-fledged economic actor, representing a significant share of the economic output in many sectors of activity. In fact, the social economy is a substantial if not *the* principal employer in certain activities such as people services. The integration of volunteer work and non-market output in the tables of national accounting data allows to quantify the contribution of the social economy to the overall wealth of a country. However, this important clarification fails to shed light on the issues of civic engagement and solidarity. Secondly, the social economy is a specific economic model that combines different resources. Research on the respective share of private resources (contributions, revenues from activities) and public resources (grants, contracts) offers among others a nuanced picture of the dependence of the social economy on public financing. Changes in public management tend to alter the mode of interaction between the social economy and the public actor (grants, calls for tender or contracts for the delegation of services, etc.). Thirdly, the social economy is a territorialized economy in order to respond to the needs of the population. This presence of the social economy on the territories, be they urban or rural, is well demonstrated by the statistics, albeit without addressing the question of the causal links.

Nonetheless, the production of social economy statistics has difficulties in expressing the full range of characteristics of this economy. Challenges regarding the internal governance of social economy organizations, related to economic democracy and collective entrepreneurship, are insufficiently taken into consideration.[23] As well, volunteerism is interpreted primarily as a resource or as free labor, thereby undermining its crucial role in the creation of social ties or the involvement of citizens in their community. Furthermore, progress is still to be made in understanding the spillover effects and structuring impacts of social economy organizations in the different sectors.

The production of social economy statistics is as much a scientific question as a political one in the sense that statistical results are decision-making tools for public and private actors. The recent introduction of legislation on the social economy in many countries[24] follows, without doubt, on the heels of the considerable effort made to identify the size and impact of this segment of the economy. This, in turn, corroborated the conceptual groundwork that defined the field and the statistical indicators, since laws identify – and thereby stabilize – the objects and standards they must comply with. Thereby, the statistical, institutional and political representations of the social economy are intimately related. Consequently, the production of social economy statistics will undoubtedly continue to progress over the course of the coming years. It remains to be seen which representations of the social economy this development will then foster.

References

ANTEAG (Associação Nacional dos Trabalhadores e Empresas de Autogestão e Participação Acionária), *Atlas da Economia Solidária no Brasil*, São Paulo, Todos os Bichos, 2009.

APRÈS-GE, "Photographie de l'économie sociale et solidaire à Genève," *Statistical Study*, Geneva, Après-Ge, 2010.

Archambault, E., *Le secteur sans but lucratif. Associations et Fondations en France*, Paris, Economica, 1996. (English version: *The Nonprofit Sector in France*, Manchester University Press, 1997).

Archambault, E., "Le travail bénévole en France et en Europe," *Revue française des affaires sociales*, No. 2, 2002, pp. 13-36.

---

[23] With the exception of a few examples, such as the study on women on the boards of directors according to which "41.1% of all board members are women, which compares to just 11.7% of board members in FTSE 100 companies and 4.9% in AIM-listed companies" (Leahy and Villeneuve-Smith, 2009: 25) or the participation in democratic governance (Brazil, 2007).

[24] Laws or decrees on the social economy exist in many parts of the world (e.g., Wallonia in Belgium (2008), Greece, Spain and Mexico (2011), Colombia and Ecuador (2012), Portugal (2013), Québec (2013), France (2014)) or are in the process of being adopted, such as in Argentina, Brazil, Luxembourg.

Archambault, E., *The American and the French Third Sectors: A Comparison, Recent Trends during the "Millennium Boom," and the Impact of the Crisis*, paper presented at the 9th ISTR Conference: Facing Crises: Challenges and Opportunities Confronting the Third Sector and Civil Society, Istanbul (Turkey), 2010.

Australian Bureau of Statistics, *Australian National Accounts: Non-Profit Institutions Satellite Account*, 2007 and 2004.

Autès M., "Les auteurs et les référentiels," in J.-N. Chopart *et al.*, *Les dynamiques de l'économie sociale et solidaire*, Paris, La Découverte, 2006, pp. 81-113.

Barea, J. and Monzón, J. L., *Manuel pour l'établissement des comptes satellites des entreprises de l'économie sociale: coopératives et mutuelles*, Liège, CIRIEC, 2006.

Bazin, C. and Male, J., *Économie sociale: bilan de l'emploi en 2010*, Paris, Recherches et Solidarités, 2011.

Belgium, *Comptes nationaux. Le compte satellite des institutions sans but lucratif*, Brussels, Institut de comptes nationaux de la Banque nationale de Belgique, 2008.

Ben Sedrine, W., Fecher, F. and Sak, B., *Comptes satellites pour les coopératives et mutuelles en Belgique. Première élaboration (SATACBEL): rapport final 2011*, Liège, CIRIEC, 2011.

Boarini, R., Johansson, A. and Mira d'Ercole, M., *Alternative Measures of Well-Being*, OECD Social, Employment and Migration Working Papers, 2006, No. 33.

Bouchard, M. J., Rousselière, D., Ferraton, C., Koenig L. and Michaud, V. *Portrait statistique de la région administrative de Montréal*, Montreal, Canada Research Chair on the Social Economy, HS-2008-01, 2008.

Bouchard, M. J., Ferraton, C., Michaud, V. and Rousselière, D. *Bases de données sur les organisations d'économie sociale, la classification des activités*, Montreal, Canada Research Chair on the Social Economy, R-2008-01, 2008.

Bovar, O., Desmotes-Mainard, M., Dormoy, C., Gasnier, L., Marcus, V., Panier, I. and Tregouët, B. *Les indicateurs de développement durable – Dossier*, Paris, INSEE, 2008.

Braibant, M., "Un outil de synthèse économique pour la politique sectorielle. Les comptes satellites," *Courrier des Statistiques*, 1994, No. 69, pp. 33-39.

Caillavet, F., "La production domestique des femmes réduit l'inégalité des revenus familiaux," *Économie et Statistique*, 1998, No. 311, pp. 75-89.

Chebroux, J.-B., "Les observatoires locaux: quelle méthodologie pour les conduire?," *Socio-logos*, Vol. 6, 2011. http://socio-logos.revues.org/2620.

Chiapello, E. and Desrosières, A., "La quantification de l'économie et la recherche en sciences sociales: paradoxes, contradictions et omissions. Le cas exemplaire de la positive accounting theory," in F. Eymard-Duvernay (ed.), *L'économie des conventions. Méthodes et résultats Tome 1*, Paris, La Découverte, 2006, pp. 297-310.

CNCRESS (Conseil National des Chambres Régionales de l'Economie Sociale), *Atlas commenté de l'économie sociale et solidaire*, Paris, Observatoire national de l'économie sociale et solidaire and Juris editions, 2012.

Comeau, M., *Étude d'impacts socio-économiques des entreprises d'insertion du Québec*, Collectif des entreprises d'insertion du Québec, Consultations Libera Mutatio, 2011.

Demoustier, D., "Le bénévolat, du militantisme au volontariat," *Revue française des affaires sociales*, 2002, Vol. 4, No. 4, pp. 97-116.

Demoustier, D. and Ramisse, M.-L., *L'emploi dans l'économie sociale et solidaire*, Domont (France), Thierry Quinqueton Editeur, 2000.

Desrosières, A., *Pour une sociologie historique de la quantification. L'argument statistique I, and Gouverner par le nombre, l'argument statistique II, two volumes*, Paris, Mines Paris-Tech, 2008.

DiMaggio, P.-J. and Powell, W.-W., "The Iron Cage Revisited: Institutional Isomorphism and Collective Rationality in Organizational Fields," *American Sociological Review*, 1983, Vol. 48, pp. 147-160.

Fecher, F., *Comptes satellites pour les coopératives, mutuelles et sociétés à finalité sociale en Belgique*, Colloque du CIRIEC-Canada au Congrès de l'ACFAS, Quebec, Université Laval, May 2013.

Gadrey, J., "Les conventions de richesse au coeur des comptabilités nationales. Anciennes et nouvelles controverses," in Eymard-Duvernay, F. (ed.), *L'économie des conventions. Méthodes et résultats, Tome 1*, Paris, La Découverte, 2006, pp. 311-324.

Gadrey, J. and Jany-Catrice, F., *Les nouveaux indicateurs de richesse*, Paris, La Découverte, 2005.

Gignac, R. and Hurteau, P., *Mesurer le progrès social, vers des alternatives au PIB*, Montreal, Institut de recherche et d'informations socio-économiques (IRIS), 2011.

Hall, M. H., Barr, C. W., Easwaramoorthy, M., Sokolowski, S. W., and Salamon, Lester M., *The Canadian Nonprofit and Voluntary Sector in Comparative Perspective*, Toronto, Imagine Canada, 2005.

Hall, M. H., de Wit, M. L., Lasby, D., McIver, D., Evers, T., Johnston, C., McAuley, J., Scott, K., Cucumel, G., Jolin, L., Nicol, R., Berdahl, L., Roach, R., Davies, I., Rowe, P., Frankel, S., Crock, K. and Murray, V., *Force vitale de la collectivité: Faits saillants de l'Enquête nationale auprès des organismes à but non lucratif et bénévoles*, Statistique Canada, 2004.

IFF Research, *A Survey of Social Enterprises Across the UK*, The Small Business Service, 2005.

Itçaina, X., Palard J. and Ségas S. (eds.), *Régimes territoriaux et développement économique*, Rennes, Presses Universitaires de Rennes, 2007.

Lahire, B., *L'esprit sociologique*, Paris, La Découverte, 2005.

Lallement, M. and Spurk, J. (eds.), *Stratégie de la comparaison internationale*, Paris, CNRS Editions, 2003.

Laville, J.-L. (ed.), *L'économie solidaire: une perspective internationale*, Paris, Desclée de Brouwer, 1994.

Laville, J.-L., Lévesque, B. and Mendell, M., "The Social Economy: Diverse Approaches and Practices in Europe and Canada," in A. Noya and E. Clarence (eds.), *The Social Economy. Building Inclusive Economies*, Paris, OECD, 2007, pp. 155-188.

Lazarsfeld, P., "Des concepts aux indices empiriques," in E. Boudon and P. Lazarsfeld (eds.), *Le vocabulaire des sciences sociales. Concepts et indices*, Paris and La Haye, Mouton, 1965, pp. 27-36.

Leahy, G. and Villeneuve-Smith, F., *State of Social Entreprise Survey 2009*, London, Social Enterprise UK, 2009.

Lévesque, B., Malo, M.-C. and Girard, J.-P., *L'ancienne et la nouvelle économie sociale, deux dynamiques, un mouvement? Le cas du Québec*, Montreal, UQAM, Chaire de coopération Guy-Bernier, 2005, No. 004.

London Economics, *Study on the impact of co-operative groups on the competitiveness of their craft and small enterprise members*, Brussels, European Commission, Enterprise and Industry, 2008.

Marée, M. et al., *Le secteur associatif en Belgique. Une analyse quantitative et qualitative*, Brussels, Fondation Roi Baudouin, 2005.

Mertens, S., *Vers un compte satellite des institutions sans but lucratif en Belgique*, PhD thesis in economics, Université de Liège, 2002.

Mertens, S. and Lefèbvre, M., "La difficile mesure du travail bénévole dans les institutions sans but lucratif," in Institut des Comptes Nationaux, *Le compte satellite des institutions sans but lucratif 2000 et 2001*, Brussels, Banque nationale de Belgique and Centre d'Economie Sociale de l'Université de Liège, 2004, pp. 1-9.

Mertens, S. and Marée, M., "The Limits of the Economic Value in Measuring the Global Performance of Social Innovation," in A. Nicholls and A. Murdock (eds.), *Social Innovation: Blurring Boundaries to Reconfigure Markets*, Palgrave Macmillan, 2012.

Prouteau, L. and Wolff, F. C., "Le travail bénévole: un essai de quantification et de valorisation," *Économie et Statistique*, 2004, No. 373, pp. 33-56.

Rousselière, D. and Bouchard, M. J., "Effets d'éviction ou de renforcement des politiques publiques à destination de l'économie sociale. Une analyse de Montréal," *Revue économique*, 2011/5, Vol. 62, pp. 941-955.

Salamon, L. M., "Putting the Civil Society Sector on the Economic Map of the World," *Annals of Public and Cooperative Economics*, 2010, Vol. 81, No. 2, pp. 167-210.

Salamon, L. M., Anheier, H., List, R., Toepler, S., Sokolowski, W. S., and Associates, *Global Civil Society. Dimensions of the Nonprofit Sector*, 2nd ed., Baltimore, Johns Hopkins Center for Civil Society Studies, 2001.

Salamon, L. M. and Dewees, S., "In Search of the Nonprofit Sector," *The American Behavioral Scientist*, 2002, Vol. 45, No. 11, pp. 1716-1740.

Salamon, L. M., and Sokolowski (eds.), *Global Civil Society*, Vol. 2, New York, Kumarian Press, 2004.

Statistics New Zealand, *Non-profit Institutions Satellite Account: 2004*, Wellington, Statistics New Zealand, 2007.

Stiglitz, J., Sen, A. and Fitoussi, J. P., *Performances économiques et progrès social, Richesse des nations et bien-être des individus*, Vol. I, Paris, Odile Jacob, 2009.

Tice, H. and members of the NPI Handbook Test Group, *Portraying the Nonprofit Sector in Official Statistics: Early Findings from NPI Satellite Accounts*, paper prepared for the 27th General Conference of the International Association for Research in Income and Wealth, Djurhamn (Sweden), August 18-24, 2002.

United Nations, *Handbook on Non-Profit Institutions in the System of National Accounts*, Department of Economic and Social Affairs, Statistics Division, Series F, No. 91, New York, 2003.

Vanoli, A., *Une histoire de la comptabilité nationale*, Paris, La Découverte, 2002.

Vienney, C., *L'économie sociale*, Paris, La Découverte, 1994.

Mapping the Field of the Social Economy
Identifying Social Economy Entities[1]

Marie J. BOUCHARD

Full professor, Université du Québec à Montréal, Canada

Paulo CRUZ FILHO

Professor, FAE Business School, Brazil

Martin ST-DENIS

Economist, MCE Conseil, Canada

Introduction

The starting point in qualifying the social economy is that all its definitions underline the primacy of the social purpose over the economic activity. This applies in particular to the empirical features that are typical of the structures and the operation of the social economy and that distinguish it from the rest of the economy. This chapter analyzes some of the most important statistical studies on the social economy conducted by researchers, academic experts, public institutions and statistical offices in various parts of the world. The resulting conceptual frameworks for producing statistics about the social economy usually establish which type of entities, legal statuses and activity sectors are excluded and identify a cluster of qualification criteria and statistical indicators of social economy organizations. Typologies of organizations can also be determined for other criteria, such as the goals and missions or the modes of financing them. A conceptual framework for qualifying social economy organizations should also allow assessing peripheral developments or trends in this field and to anticipate its

---

[1] We wish to thank Damien Rousselière for his comments and suggestions in the preparation of this chapter.

progress (e.g., the recognition of new organizations as belonging to the social economy, and their integration).

The first task in any production of statistics is to define the "object" or the "beings" to be measured (Desrosières, 1993), namely, by defining the rules for building the statistical population. This is what will allow to determine the components and the boundaries of the statistical population in an approach we call "qualification." Such an approach will also allow to identify the different types of entities that belong to a statistical field under study. Since the social economy is, as any field from the social sciences, subject to change, the qualification system should be designed to anticipate its own evolution.

In the case of the social economy, one of the difficulties is that, apart from a few exceptions,[2] the national statistics systems have no markers for clearly identifying or distinguishing this subset of the economy. This is due in part to the fact that in most countries the social economy is still poorly codified within public policy. Another difficulty with the social economy is that, in addition to the various names this economy goes by depending on the region (e.g., social economy, solidarity economy, popular economy), it comprises entities that are qualified by their organizational structures and their modes of operation, or by the values that drive them, all of which are criteria that are not always easy to observe with a view to producing statistics. Moreover, a third difficulty is the permeable nature of the boundaries of the social economy, which is often seen to consist of a "hard core" and "peripheral" (Desroche, 1983) or "hybrid" (Spear, 2011) components that are surrounded by "porous borders." Finally, emerging concepts such as that of the social enterprise as well as the need for international comparability have prompted a search for the boundaries of the social economy beyond the statutory schemes traditionally associated with it, such as the cooperatives, mutual societies, associations and foundations.[3]

Despite these difficulties, statistics on the social economy have been produced in different national contexts for some years. This chapter shows the ways in which these statistical portraits qualify the statistical population of the social economy, and how these ways vary depending on the objectives of the statistical studies and on data availability. We explore these issues by following a sequence, following the methodological

---

[2] As in the case of France (see the chapter by Demoustier *et al.*).

[3] See in particular the concept of social enterprise used by the European Commission: http://ec.europa.eu/enterprise/policies/sme/promoting-entrepreneurship/social-economy/social-enterprises/index_en.htm (accessed 16 July 2014).

steps so as to arrive at coherent statistical data. The first step is choosing a definition that is operational for statistical purposes and that concurs with the social construct of the social economy. Another step involves choosing the qualification criteria based on this operational definition. These criteria then serve to screen the entities that will form the population under study. The chapter concludes with a discussion of some of the issues that arise during the building of the statistical definition of the social economy.

This chapter analyzes some of the most important statistical studies on the social economy conducted by researchers, academic experts, public institutions and statistical offices between 2005 and 2012. We base our discussion on 15 statistical portraits produced in various countries on the social economy, on the social and solidarity-based economy, or on social enterprises.[4] The portraits were drawn from ECO-SOC INFO, a bulletin on social economy research issued by the Canada Research Chair on the Social Economy,[5] as well as a complementary study conducted primarily on the Internet.[6]

The geographical scope of the selected portraits is varied. While some portraits focus on regions of a country (Elson and Hall, 2010; Gesky, 2011) or specific administrative regions or cities (Bouchard *et al.*, 2008; Pellet, 2009), others are national or even international in scope (national: ANTEAG, 2009; Barraket *et al.*, 2010; Clarke and Eustace, 2009; IFF Research, 2005; INE, 2012; Mecherkany 2010; Monzón Campos, 2010; Observatoire national de l'économie sociale et solidaire, 2009; international: Chaves and Monzón Campos, 2007, 2012).

---

[4]   While recognizing that these terms cover different scopes, we use the term "social economy" for purposes of simplification.

[5]   The issues of the newsletter can be found on the website: http://www.chaire.ecosoc.uqam.ca.

[6]   This study was first conducted in 2010-2011 in order to produce, for the Institut de la statistique du Québec, a conceptual framework for defining the statistical population of the Québec social economy (see Bouchard, Cruz Filho and St-Denis, 2011). It is based on the websites of various social economy research centers whose primary language is English, French, Spanish or Portuguese, as well as on search engines, likewise in these four languages, using keywords combining synonyms of "portrait" (study, research, etc.) and concepts related to the social economy. It is important to note that the analysis presented here does not claim to offer exhaustive statistical portraits on the social economy, but aims rather to highlight the key terms of qualification of the social economy for statistical purposes.

1. From social construct to statistical reality

The statistical portraits retained for our analysis, although covering broad fields and applying different methods of qualification, nevertheless show three main similarities. One, they establish an operational definition of the social economy. Two, they determine qualifying criteria on the basis of this definition. And three, they establish filters that allow to distinguish between the entities of a field, including between those who belong to the statistical population and those who do not. The first two steps (definition and criteria) are presented hereafter. The third step (filters) is the subject of the second part of the chapter.

1.1. An operational definition of the social economy

The development of a "statistical definition" of the social economy that is reliable and relevant to policy makers and to the actors of the movement should ideally be based on the definition that is currently in usage and that is institutionalized (in the sense of being generalized) in society. A very inclusive definition allows to cover other concurrent definitions, which are essentially subsets of that broader definition. This allows to situate oneself ahead of the debates while identifying the different components of the sets (e.g., cooperatives, non-profit sector, non-statutory social enterprises). The portraits we studied concern the fields of the *social economy* (Bouchard *et al.*, 2008; Constantinescu, 2011; Elson and Hall, 2010; Monzón Campos, 2010), the *social and solidarity-based economy* (Observatoire national de l'économie sociale et solidaire, 2009; Pellet, 2009), the *solidarity economy* (ANTEAG, 2009) and *social enterprises* (Barraket *et al.*, 2010; Clarke and Eustace, 2009; IFF Research, 2005; Leahy and Villeneuve; Mecherkany, 2010). The definitions used by these portraits are those conveyed by international (INE, 2012; Monzón Campos, 2010) or national (ANTEAG, 2009; Bouchard *et al.*, 2008; Geski, 2011) institutions and groups and are sometimes inscribed in the law (e.g. Spain).

Whether in the case of the social economy or any other social construction, the concepts seek to represent as closely as possible the characteristics of a social construct for which there is a consensus or near-consensus. However, such a social construct does not necessarily lend itself for statistical qualification, based as it is on a set of goals, principles and values, all of which are notions that are difficult to observe empirically.

In addition, the social economy, while part of a broader global movement, remains fundamentally rooted in the needs of the people and the communities in which it is embedded. Subsequently, its reality may differ significantly from one national context to another (Bouchard, Cruz Filho and St-Denis, 2011). Therefore, the definitions used in statistical portraits must be both general enough to show the link of the social economy to an international movement and specific enough to accurately reflect its local originality. However, this calls for the identification of empirical, stable and readily observable indicators, which in turn explains the importance of finding an operational definition within a qualification process.

The construction of operational definitions is common practice in social statistics. For example, it is this that has allowed for the emergence, in Canada, of statistics on visible minorities (Beaud and Prévost, 1999), of statistics on small and medium enterprises in the member states of the European Union (European Commission, 2003) and of satellite accounts on nonprofit institutions in many countries of the world (United Nations, 2003).

As the social economy concept is based on principles, values and rules (Defourny and Develtere, 1999; Draperi, 2007; Vienney, 1980, 1994), an operational definition of the social economy must translate these elements into qualification criteria from which to build observable empirical indicators. The construction of this definition involves mediating between the cost of measuring an element and the contribution of that element to the area studied.

1.2. Identification of qualification criteria

For practical reasons, the qualification should be based on empirically observable attributes that allow to distinguish the entities that are inside the field from those that are outside. Statistical work done on the social economy requires first to identify the economic sectors and activity sectors that are most likely to contain those entities, and then to sort them according to distinctive observable characteristics. A first exercise is to examine how the definition of the social economy matches the major classification systems used by the statistical agencies.

The logic model for identifying the entities that make up the social economy generally includes three main groups of tasks: 1) the identification of entities in the economic sectors and the activity sectors most likely to contain social economy organizations; 2) the selection of entities by their legal status; 3) the sorting out of entities that match a set of qualification criteria in order to identify, from entities that either have or do not have the legal status of social economy organizations, those that belong to the social economy according to the institutional definition used.

In the system of national accounts, social economy organizations belong mainly to the sectors of non-financial firms, incorporated financial firms and non-profit institutions serving households.[7] In many countries,

---

[7]   See the chapter by Édith Archambault.

73

for historical reasons, certain activity sectors – professional associations, employer groups, political parties, religious organizations and unions – are generally excluded from the definition of the social economy and from the statistical portraits (Bouchard, Cruz Filho and St-Denis, 2011; INSEE, 2011). These excluded activity sectors can usually be easily identified in the national classification systems (European Commission, 2008; United Nations, 2009; Statistics Canada, 2012).

The legal statuses that are generally recognized as belonging to the social economy are those of cooperatives, mutual societies and associations (or non-profit organizations) and sometimes also foundations (as in France).

The criteria used to define the social economy vary according to the operational definition used in the statistical portraits. However, five criteria have emerged as key elements:

- The social mission, a principle that overrides all other criteria and the manifestation of which is ensured through the verification of other criteria.

- The limited or prohibited distribution of surpluses, which manifests mainly in an organization's legal personality (cooperative, mutual societies, association, foundation).

- The organized production of goods or services, which can be verified, among others, through the reported sales of goods or services or the presence of salaried employees or volunteers. Thus, organized production may be entirely subsidized or realized by volunteers, and subsequently is not always synonymous with commercial activity, even though market-based organizations do meet this criterion.

- The criteria for autonomy and independence refer to the "control which an entity (individual, group, instance) has of its interdependencies with other beings, measures or objects" (Eme, 2006, p. 173). While some definitions refer only to independence vis-à-vis the state, others exclude any affiliation with entities that do not belong to the social economy.

- Democratic governance involves the right of stakeholders (users, workers, members) to oversee decisions in the organization. This criterion is defined in several ways depending on the region and the realities encountered. In Brazil, for example, the national survey on the solidarity economy (ANTEAG, 2009) views democracy on the basis of the self-management principle. According to this principle, organizations are formed and run by workers who decide collectively on the management of activities and the allocation of resources.

2. The effect of screening methods on qualification

Screening is the method by which the indicators arising from qualification criteria are applied. It should be noted that the qualification criteria do not all require the same degree of verification. Some criteria are more easily identifiable than others, depending on the indicators mobilized. Thus, the distribution of benefits is an observable criterion. By contrast, democracy is a latent and unobservable phenomenon, although it can be approximated with certain observable variables. By nature, this phenomenon is therefore subject to greater measurement error. For each qualifying criterion, several indicators can be identified, based on the possibilities and means available to those responsible for the study. In statistical portraits of the social economy, the most obvious and common indicators are belonging (or not belonging) to an activity sector and legal status.

By successive screening, the qualification criteria can be sequenced so as to obtain, after each screening, an increasingly accurate portrait of the population studied. At that point, the arbitration between the quality and the costs of the qualification comes into play. Each filter consists of indicators and a screening method. The indicator is the variable that is observed for a criterion, while the screening method is the way in which this variable is measured. Figure 1 shows, in a simplified way, an example of the application of successive filters.

Figure 1. *Application of successive filters*

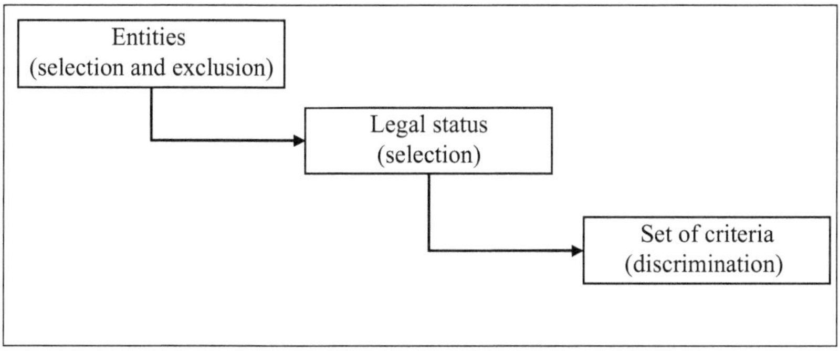

Source: Bouchard, Cruz Filho and St-Denis, 2011, p. 26

In Figure 1, the first filter applied concerns the inclusion (or exclusion) of entities in the economic sectors (national accounting systems) and activity sectors (classification systems of industries or activities) most likely to contain social economy organizations. In the latter case, the use of administrative databases allows removing households, organizations in public administration and entities excluded for historical reasons. The second filter concerns the selection of organizations by the indicator of their legal status. For illustrative purposes, the portrait of the social and solidarity-based economy in France includes cooperatives, mutual societies, associations and foundations. In the third filter, a set of qualification criteria allow for a finer screening of the organizations selected by the two preceding filters.

The analysis of statistical portraits indicates that screening methods can be applied in at least three ways, namely: selection based on administrative data; the validation of one or more lists of groupings of a sector (or control organizations); and the use of screening questions in a qualification questionnaire.

2.1. The quality of the screening

The screening method chosen in the different studies results from a trade-off between the quality of the qualification and the costs of obtaining information. Although "even now, survey and quality are vague concepts" (Lyberg, 2012, p. 108), the quality of a statistical scope "can be measured and controlled by means of degree of adherence to specifications and requirements" (Lyberg, 2012, p. 114). For the purposes of the analysis, we consider that the quality of the qualification can be divided into five different issues: the coverage of the study; the integrity (vitality) of the qualified organizations; data availability; comparability of the study; and sustainability (continuity) of the study, which refers to its potential for replication.[8]

The coverage of the study refers to the degree to which the qualified organizations match or reflect the initial definition. It identifies whether the screening method can handle exceptions and exclusions that are specific to the chosen definition and questions the treatment of borderline cases (entities with some but not all of the qualifying criteria); it helps to put qualification issues into perspective with the law of large numbers (i.e., if the sample under study is large, the impact of the particular characteristics on the validity of the results will be low);[9] it addresses the question of informal organizations; and it recognizes the social economy as a group of families of organizations on which there is not always consensus.

The integrity or vitality of the screening frame allows to situate the statistical portrait in time. Integrity, as a characteristic of a database, or of the statistics developed therefrom, pertains to the quality of maintaining information that is up to date and accurate when being used or accessed.

---

[8] It should be added that portraits should, according to the quality requirements for any statistical survey, mention potential measurement errors and their sources (Lyberg, 2012).

[9] This only applies if the objective is to determine averages or typologies. It does not apply when seeking to emphasize the statistical weight.

The integrity is affected when the statistics include a large number of organizations that have closed down or that have not provided information for a long time. It may also involve the omission of new organizations. Data that are outdated reflect a picture of the past (although some elements may still be true even when they have not been updated). Also, the retention of organizations that have ceased their activities swells the population artificially, calling for a validation of the data integrity. The inclusion of emerging organizations is also problematical.

The availability of data concerns the limitations of a screening method. How much information might a screening method allow us to obtain? Which qualification criteria might we fulfill?

The screening method also determines the comparability of studies and the comparability of results (which also depend on homogeneous methodology). Further, comparability indicates, among other aspects, whether regional results are comparable with national results and allows to contextualize a study in relation to other statistical realities.

Sustainability or continuity indicates whether the screening method allows for a longitudinal tracking of the qualified population. A study's sustainability is shaped by several factors, among them the intertemporal comparability of the qualification. This factor also concerns stability and changes in scope due to the inclusion or exclusion of borderline cases.

2.2. Screening methods

The decision to use primary data obtained through surveys or the use of secondary data obtained from administrative sources orients the operationalization of the qualification and, therefore, the choice of methodology of the study. In some cases, recourse to databases that are made available by the national statistical institutes allows for the adoption of statistical indicators, such as legal status or the area of activity, that identify social economy organizations in one go. Once the organizations are selected, it is possible to survey them or to draw a portrait of them by extractions of the administrative databases. The screening operation can also be performed by sectoral associations, which compile lists of their members. The screening can likewise be performed through the questions posed in a survey.

2.2.1. The selection based on administrative data

The digitization of administrative data from the 1960s and 1970s, combined with technological advances in recent years, has allowed for the maintenance and use of large databases by public institutions. The nature of administrative data is shaped by the sources they come from, such as statutory tax returns, for example. Moreover, statistical institutes enrich these administrative data with mandatory or voluntary surveys (Statistics Canada, 2010).

The use of administrative data in drawing a statistical portrait is conditional on the presence, on the territory where the study is conducted, of a statistical institute with the necessary means to process that data. It also requires that these administrative data be considered by the statistical agencies as a potential storehouse of information. The same applies to the different public and parapublic agencies that often possess and analyze considerable amounts of information on certain sectors of the social economy. These can contribute to the building of partial portraits of the social economy.

In addition, it requires a certain degree of institutionalization of the components of the social economy, often through the form of the legal statuses. Moreover, to allow for processing at a lower cost, these data must be digitized. Normally, administrative data are available to statistical offices or with special authorization. However, while these conditions are often met in Western countries, this is not the case in many emerging economies (Fioretti, 2011, p. 8).

Two indicators that are frequently operationalized using administrative data are legal status and sector of activity (Barea and Monzón Campos, 2006; Bouchard, Cruz Filho and St-Denis, 2011; IFF Research, 2005; Monzón Campos, 2010; Observatoire national de l'économie sociale et solidaire, 2009; United Nations, 2003). The *Atlas de l'économie sociale et solidaire en France et dans les régions* also identifies employing organizations, using administrative data (Observatoire national de l'économie sociale et solidaire, 2009).

The screening method using administrative data differentiates between three states of qualification, designated by the letters A, B and C in Figure 2.

Figure 2. *Models of statistical portraits that can be built using administrative databases*

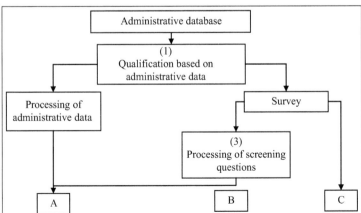

Administrative data can be used at two levels: the construction of a parent population to be used at a later point for a survey (states B and C) or the building of a statistical portrait or satellite account with the available data, thereby foregoing the need to produce a survey (state A). We focus here on the use of administrative data for the purpose of performing qualifications rather than that of obtaining results (filter 1).

Coverage

When using administrative data, the completeness of databases becomes a qualifying issue, since a selection based on certain criteria, such as size of the organization or activity sector, could have an impact on the entire set of organizations belonging to the statistical portrait. Indeed, several types of administrative databases are usable for the qualification of the social economy. Some are very large and exhaustive, as is the case with databases associated with national accounts, while others are very specific to the reality of the concept studied (Statistics Canada, 2007a).

When national accounts (associated with tax data) are used for screening, they exclude the informal organizations and those who do not engage in commercial activities (Statistics Canada, 2010, p. 3). Since administrative data are based on organizations that are indexed, such a method may not cover the informal sector of the economy – which may nevertheless constitute a significant segment of the social economy, as seen in the case of Brazil, for example (ANTEAG, 2009).[10] Also, organizations that are still very emerging are excluded, since statistical institutions do not systematically identify them. The same limit applies to organizations that are not subject to certain obligations, such as associations that are exempt from paying certain taxes in certain contexts. In addition, in some sectors, such as agriculture in France, various public institutions are involved in the collection of data, resulting in a very dispersed distribution of the available administrative data. Apart from the cases mentioned above, the combination of "Generally Accepted Accounting Principles (GAAP) which are used in the business world, to concepts [of a national accounting system], the [administrative data represent] the business world realities in a format that meets our statistical needs" (Statistics Canada, 2010).

Integrity

The integrity of administrative data depends on the effort invested in that aspect, hence the importance of the credibility of the statistical institute in charge of this matter. This integrity is ensured through monitoring (Statistics Canada, 2010) or by consulting surveys conducted by the statistical institute (INSEE, 2013; Statistics Canada, 2010), allowing to remove

---

[10] See the chapter by Luiz Inácio Gaiger.

the organizations that have ceased their operations. Statistics Canada notes, however, that "it is easier to deal with businesses that are dead during estimation (using domain estimation) than it is to deal with businesses thought to be dead, but which are alive (leading to under coverage)" (Hunsberger, Beaucage and Pursey, 2005).[11]

Availability of data

The variables in the administrative data often allow for broad divisions. Legal status, activity sector or employer status are all indicators that are generally verifiable in the administrative data (Bouchard, Cruz Filho and St-Denis, 2011). However, certain indicators allowing to cover more detailed qualification criteria are not available in the administrative data and require conducting a survey, as was the case in the United Kingdom (IFF Research, 2005).[12] For example, if the legal status is insufficient, screening questions meeting the criteria of democratic governance or autonomy may be asked (Bouchard, Cruz Filho and St-Denis, 2011).

Territorial comparability

Since 1953, the Statistics Division of the United Nations has been publishing the System of National Accounts, periodically revised (European Commission *et al.*, 2009). This document establishes standards for statistical institutes around the world for the building of national accounts, which in turn serve as the bases for national economic studies. Moreover, statistical offices use these standards for their respective administrative data, which ensures considerable comparability of indicators resulting from that data across similar studies at the international level.

Sustainability and potential for replication

The intertemporal qualification on the basis of administrative data is subject to variations in the definitions of the variables. One example might be the updating of classification systems, as was the case with the North American Industry Classification System (NAICS) in 2007 (Statistics

---

[11] Statistics Canada analysts nevertheless noted an ambiguity regarding the maintenance of such databases: Although the treatment of organizations that ceased operations is easy to accomplish in the estimates of statistics by area (analysis of a specific segment of the population to identify the mortality that is specific to it), the possibility that a proportion of them is still active leads to undercounting, which still remains a methodological challenge (Hunsberger, Beaucage and Pursey, 2005). There are some models that allow to assess the gain in accuracy versus the cost of the collection. See for example Kalton (2001) on the analysis of rare populations. Such models are increasingly used. Nevertheless, social economy portraits are still produced in a very empirical manner, which can lead to poor estimates.

[12] See the chapter by Roger Spear.

Canada, 2007b) and in 2012 (Statistics Canada, 2012). These changes may affect certain indicators that were previously used.

In sum, the use of administrative data allows for a broad and inclusive qualification using only few criteria and indicators. The resulting qualification allows to build a list likely to contain social economy organizations (Bouchard, Cruz Filho and St-Denis, 2011). This screening method is preferred by the national portraits (Monzón Campos, 2010; Observatoire national de l'économie sociale et solidaire, 2009) and the satellite accounts (Barea and Monzón Campos, 2006; United Nations, 2003).

2.2.2. The validation of one or more lists of groupings from the sector

Outside of statistical institutes, it becomes much more difficult to qualify organizations based on administrative data. The portraits of the social economy done outside the statistical agencies offices therefore require the establishment of a target population allowing to conduct a sample survey. Figure 3 shows the different qualification patterns found in the studied portraits.

Figure 3. *Models of statistical portraits that use lists of social economy groups*

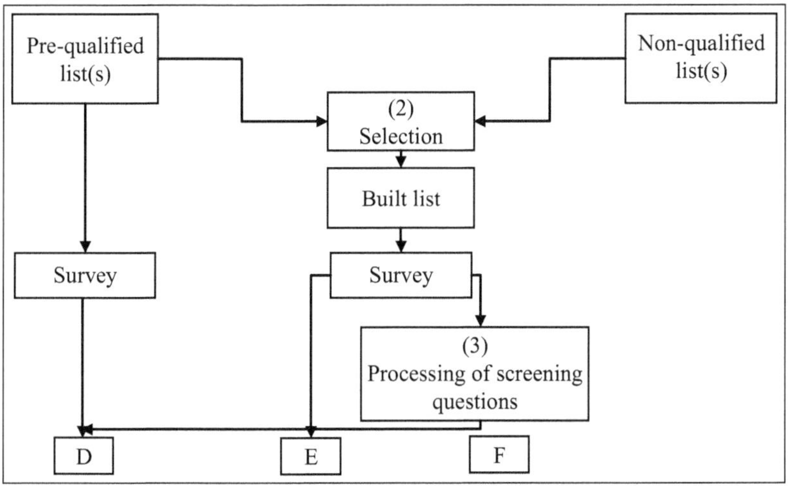

In Ireland, the United Kingdom, Sweden as well as Geneva (Clarke and Eustace, 2009; Leahy and Villeneuve, 2009; Mecherkany, 2010; Pellet, 2009), the step of validating the population was bypassed by the use of only one list, namely the list considered to be exhaustive with regard to the operational definition chosen by the authors (case D). For the studies conducted in Canada and Australia, the target population was built from a crossing of lists issued from different groups or sector associations (ANTEAG, 2009; Barraket *et al.*, 2010; Bouchard *et al.*, 2008; Elson and Hall, 2010) (cases E and F). The latter approach is conditional on the existence of social economy organizations that are recognized in the territory under study. This way of screening allows for a more exclusive focus on organizations that are within the social economy networks. In addition to the list-crossing method, one may add identification by word of mouth (ANTEAG, 2009; Barraket *et al.*, 2010).

Coverage

Without completeness of coverage, the estimates that are realized and the results of the portrait are potentially biased (Bouchard *et al.*, 2008, p. 26). When it comes to crossing lists, these may be pre-qualified, such as those coming from federations or associations of social economy organizations; or they may be non-qualified, in which case they require additional validation work (Bouchard *et al.*, 2008, p. 27). Studies based on only one list of organizations cannot guarantee a completeness of coverage, since data validation by triangulation becomes impossible.[13] However, the costs of building the survey frame are much lower for such studies.

Integrity

The use of lists from associations can pose a significant challenge to the integrity. Among other aspects, the objectives for building these differed from list to list.[14] The lists are not necessarily up to date, as the registration onto and deregistration from these lists are often not mandatory. Crossing lists from associations requires the step of validating the activity and the belonging to the sector (Elson and Hall, 2010, p. 17).

Data availability

The information in lists of associations and groups may be heterogeneous from one list to the next. The study on Montreal yielded some information, such as the legal status or activity sector, which was then validated through the survey or through publicly available directories

---

[13] The challenge is different depending on whether the goal is to make a typology of organizations or a count. When making a count, the challenge is considerable, whereas when seeking to establish a typology, problems arise only if the non-presence on the list is due to random factors (rather than characteristics specific to the organizations).

[14] In the case of organizations belonging to a federation, for example, whose mandate is to appear as representative as possible, the presence of "phantoms" is more likely.

provided by government institutions (Bouchard *et al.*, 2008). However, this type information is not available in all legal systems (Fioretti, 2011). In those cases, it then becomes necessary to use screening questions to qualify organizations.

Territorial comparability

Because the lists of organizations are provided by the social economy networks (Bouchard *et al.*, 2008, p. 26), this method is more likely to present regional disparities.[15]

Sustainability and potential for replication of the study

Replicating a study based on sectoral federations lists involves starting from scratch, every time, when seeking to identify the new organizations, which becomes a very costly procedure when crossing of series of lists. As for organizations that have ceased their activities, it is possible to contact them to verify whether they are still active in the field.

2.2.3. The use of screening questions

In the two previous screening methods, the qualification of organizations is done prior to the surveys, *ex ante*. However, when the portraits resort to screening questions for the qualification, the screening method is *ex post*. In those cases, the survey frame is constructed starting from one of the two screening methods presented above, in addition to the administration of screening questions in the surveys to the end of qualifying the organizations (ANTEAG, 2009; Barraket *et al.*, 2010; Elson and Hall, 2010; IFF Research, 2005). In Figures 2 and 3, this pertains to results B and F. In this way, a survey can allow to filter organizations on the basis of a set of criteria that are expressed in the form of screening questions (Bouchard, Cruz Filho and St-Denis, 2011). This screening method is not used in all the surveys we analyzed. It allows to validate the more precise indicators and to refine the qualification; however, the application of this method constitutes an additional cost. The use of a large and inclusive sample frame is needed to use this screening method. The qualification criteria covered by the screening questions in the analyzed studies were the social mission and the level of market income (IFF Research, 2005). In the case of the survey in Brazil, the question concerned the legal nature of the surveyed organization (ANTEAG, 2009).

---

[15] The issue of representativity may nevertheless be studied on a sub-population starting from a "golden standard" (reliable external information), which then allows for assumptions about the non-coverage, which could in turn be applied to the whole population.

Coverage

The use of screening questions allows for a broad and inclusive coverage. The completeness of the coverage is determined by the screening methods that are applied *ex ante*. Screening methods allow to refine the qualification and to identify and eliminate organizations that do not belong to the field, which could not be detectable otherwise.

Integrity

The responding organizations are all "alive." With regard to the non-return issue, a distinction must be made between a non-return and a non-response. The former is the incapability to communicate with an organization, whereas the latter is a refusal to respond. In addition, some organizations are not in the position to respond because they have left the field (geographical, sectoral) under study or because the studied base has anomalies (e.g., wrong addresses). In this case, the purpose of a survey would be to find out why there was no return. A well-validated *ex ante* screening will result in fewer non-return cases associated with these situations. However, non-returns may also depend on the specific field under study and on the time that has lapsed between the constitution of the sample frame and the administration of the survey.

Availability of data

The screening questions allow for the validation of indicators that are not necessarily available otherwise. In addition, this refinement of the qualification applies not only to the respondents but also to the administration of the screening questions. The latter relies on the cooperation of the respondents, who are assumed to be familiar with the notion of social economy, and their understanding of the questionnaire. This explains the risk of an essentialist approach to social economy: If certain questions are administered only to social economy organizations, then we cannot directly deduce therefrom that the answers are unique to the field. The inclusive approach then provides validity to deal with this problem.

Territorial comparability and sustainability

The same screening questions can be used or adapted from one territory to another and from one period to another. However, it must be ensured that organizations from different territories are comparable and that they are alive, through the *ex ante* application of selection criteria.

Conclusion

A statistical portrait is intended to represent the scope of a phenomenon, to highlight its main components and their relative importance, to document some of its branches or sub-sectors, to follow its evolution[16] over time, and, if possible, to allow for comparison with other phenomena. Because it still is weakly codified in the national statistics, and because the social economy is a composite phenomenon, three steps are usually required to qualify a statistical population: the establishment of an operational definition; the identification of its variation with regard to qualification criteria; and the application of those criteria by filters that allow to distinguish the entities belonging to the field from those outside of the field. The application of the qualification can be done according to various methods, subsequent to which a population is established *ex ante* (through the use of administrative data or already established lists) or *ex post* (by administering screening questions to surveyed entities). At each stage and for each of these methods, decisions must be made that require a tradeoff between the cost and the quality of data. The issue of replicability should also be mentioned, as not all portraits provide information on the applied quality procedures and criteria.

However, a number of issues generally arise, and it would be imprudent to not mention them. The first challenge of realizing a statistical portrait is to establish an operational definition that is concordant with the concept of the social economy conveyed by the local stakeholders and the public policies. The qualification criteria established from this operational definition help determining the boundaries of the area under study. A broad and inclusive definition allows being ahead of the debates that occur when the definitions are not institutionalized, while allowing to identify the different subsets (e.g., cooperatives, non-profit sector, non-statutory social enterprises). In return, all the qualification work contributes to the "statistical hardening"[17] of the social economy. If what we measure defines what we seek (and vice versa) (Stiglitz, Sen and Fitoussi, 2009), then the production and implementation of a conceptual framework for the qualification of the social economy is likely to have a significant impact on how the social economy is perceived and, consequently, on the design, implementation and evaluation of policies concerning the

---

[16] While ensuring maximum temporal coherence, in order to allow for a longitudinal follow-up.

[17] The "process by which, after a series of interventions (designation of a spokesperson, advocacy for the defense of a group, journalistic or scholarly surveys, etc.) and formatting procedures (establishment of the definition, cut-points, equivalence classes, standardization of concepts and questions, repeated administration of surveys, etc.), phenomena marked by fuzziness, ambiguity and multiplicity of judgments are accorded a degree of univocity that authorizes their delimitation and their inventory" (Beaud and Prévost, 1999 p. 3-4, our translation).

social economy.[18] Applying successive filters performs the qualification of the target population. These filters are composed of statistical indicators (variables on which the filter is applied) and screening methods (the data source that allows for the validation of the indicator). A study using data from the Business Register of Statistics Canada[19] concluded that legal status is a primary effective filter for identifying the social economy, having found only very few businesses without a status that feature the characteristics of the social economy (McDougall, 2007). The choice of filters is a tradeoff between the quality of the qualification and the costs associated with their application.

Each stage of the qualification has an impact on the content of a statistical portrait, which explains the diversity of statistical portraits. Some survey administrators opt for a large population yet also more limited qualification criteria, which allows to exploit the law of large numbers with regard to qualification errors. However, others, operating with a more precise definition and often having to make due with few observations, cannot have organizations outside of the field without biasing the results, and therefore opt for a more refined qualification method.

The issue of intertemporal and interterritorial comparability of studies is based on the homogeneity of ways of doing. The establishment of a national observatory, as in France, allows to capture the creation and cessation of social economy organizations and facilitates population monitoring. A better adapted legal definition or a label also allows to identify, at a lower cost to the researcher, the organizations belonging to the field of study.

Finally, using multiple approaches rather than only one for the qualification allows to achieve a better quality, but is subject to higher costs. The choice of a qualification method therefore depends on the objectives of the study. For example, identifying the boundaries of the social economy within an economy requires less refinement in the qualification than does the measurement of the impact of the social economy on a restricted area. A good portrait should be able to explain the advantages and disadvantages of each of the choices that were made.

---

[18] In that regard, it is significant to note that the interest in the statistical monitoring of the social economy coincides with the adoption of laws on the social economy in a number of countries. In some cases, such as France, statistical counts – and thereby the qualification – preceded the adoption of a definition in a law (*Projet de loi relatif à l'économie sociale et solidaire*, July 3, 2014), while in other cases, such as Québec, the adoption of a law (Bill 27, *Loi sur l'économie sociale*, adopted on Oct. 10, 2013) served to define the social economy before the statistical monitoring was conducted across the entire territory.

[19] http://www23.statcan.gc.ca/imdb/p2SV_f.pl?Function=getSurvey&SDDS=1105.

References

ANTEAG (Associação Nacional dos Trabalhadores e Empresas de Autogestão e Participação Acionária), *Atlas da Economia Solidária no Brasil*, São Paulo, Todos os Bichos, 2009.

Barea, J. and Monzón, J. L., *Manuel pour l'établissement des comptes satellites des entreprises de l'économie sociale: coopératives et mutuelles*, Liège, CIRIEC, 2006.

Barraket, J. (ed.), *Finding Australia's Social Enteprise Sector: Final Report*, Queensland, Australia, University of Technology, Australian Centre for Philanthropy and Nonprofit Studies, 2010.

Beaud, J.-P. and Prévost, J.-G., *L'ancrage statistique des identités: les minorités visibles dans le recensement canadien*, Montreal, CIRST, UQAM, 1999.

Bouchard, M. J., Cruz Filho, P. and St-Denis, M., *Cadre conceptuel pour définir la population statistique de l'économie sociale au Québec*, Cahiers de la Chaire de recherche du Canada en économie Sociale, R-2011-01, Montreal, Canada Research Chair on the Social Economy/CRISES, 2011.

Bouchard, M. J., Rousselière, D., Ferraton, C., Koenig, L. and Michaud, V., *Portrait statistique de l'économie sociale de la région de Montréal*, Hors-série no HS-2008-1, Montreal, Canada Research Chair on the Social Economy, 2008.

Chaves, R. and Monzón, J. L., *L'économie sociale dans l'Union européenne. Résumé du rapport d'information élaboré pour le Comité économique et social européen*, Working Paper CIRIEC No. 2008/01, Liège, CIRIEC, 2007.

——, *The Social Economy in the European Union. Report Drawn up for the European Economic and Social Committee by the International Centre of Research and Information on the Public, Social and Cooperative Economy (CIRIEC)*, Brussels, European Economic and Social Committe, 2012.

Clarke, A., Eustace, A. and Eustace Patterson Ltd, *Exploring Social Enterprise in Nine Areas in Ireland*, n.p., Report commissioned by PLANET, 2009.

CNCRES (Conseil National des Chambres Régionales de l'Economie Sociale) and Observatoire national de l'ESS, *Atlas de l'économie sociale et solidaire en France et en régions*, Paris, 2009.

Constantinescu, S. (ed.), *Atlasul Economiei Sociale*, Bucharest, Institutul de Economie Sociala, 2011.

Defourny, J. and Develtere, P., "Origines et concours de l'économie sociale au Nord et au Sud," in J. Defourny, P. Develtere and B. Fonteneau, *L'économie sociale au Nord et au Sud*, Paris, Brussels, De Boeck & Larcier, 1999, pp. 25-56.

Desroche, H., *Pour un traité d'économie sociale*, Paris, Coopérative d'information et d'édition mutualiste, 1983.

Desrosières, A., *La politique des grands nombres, histoire de la raison statistique*, Paris, La Découverte, 1993.

Draperi, J.-F., *Comprendre l'économie sociale, fondements et enjeux*, Paris, Dunod, 2007.

Elson, P. R. and Hall, P., *Strength, Size, Scope: A Survey of Social Enterprises in Alberta and British Columbia*, Port Alberni, The BC-Alberta Social Economy Research Alliance, 2010.

Eme, B., "La question de l'autonomie de l'économie sociale et solidaire par rapport à la sphère publique," in J.-N. Chopart, G. Neyret and D. Rault, *Les dynamiques de l'économie sociale et solidaire*, Paris, La Découverte, 2006, pp. 171-203.

European Commission, *Commission Recommendation of 6 May 2003 concerning the definition of micro, small and medium-sized enterprises*, 2003/361/EC, Official Journal of the European Union, Brussels, 2003.

——, *NACE Rev. 2. Statistical Classification of Economic Activities in the European Community*, Eurostat, Methodologies and Working Papers, Luxembourg, Office for Official Publications of the European Communities, 2008.

European Commission *et al.*, *System of National Accounts 2008*, New York, United Nations, 2009.

Fioretti, M., *Open Data: Emerging Trends, Issues and Best Practices*, Pisa, Laboratory of Economics and Management, 2011.

IFF Research, *A Survey of Social Enterprises Across the UK. Research Report prepared for The Small Business Service (SBS) by IFF Research Ltd*, London, 2005.

INE (Instituto Nacional de Estatística), *Conta Satélite da Economia Social*, Lisbon, 2012.

INSEE (Institut national de la statistique et des études économiques), *B. Liste des activités soustraites de la sélection sur les catégories juridiques*, Paris, 2012.

——, *Le contenu de la base SIRENE*, Paris, 2013.

Kalton, G., *Practical Methods for Sampling Rare and Mobile Populations*, Proceedings of the Annual Meeting of the Americal Statistical Association, Alexandria, VA, Americal Statistical Association, 2001, pp. 5-9.

Leahy, G. and Villeneuve, F., *State of Social Enterprises Survey 2009*, London, Social Enterprise Coalition, 2009.

Lyberg, L., "Survey Quality," *Survey Methodology*, Vol. 38, No. 2, Catalogue No. 12-001-X, Ottawa, Statistics Canada, 2012, pp. 107-130.

McDougall, B., *Results of the 2006 Feasability Study on the For-Profit Segment of the Community Sector*, Ottawa, Human Resource and Skill Development Canada, 2007.

Mecherkany, R., *Social Entrepreneurship in Sweden. Government Support and Innovation*, Master's thesis at the Department of Transport and Economics, Royal Institute of Technology (KTH), Stockholm, 2010.

Monzón, J. L., *Las Grandes Cifras de la Economía Social en España*, Valencia, CIRIEC-España, 2010.

Pellet, T., *Étude statistique. Photographie de l'économie sociale et solidaire à Genève*, Geneva, Chambre de l'économie sociale et solidaire, 2009.

Pérez de Uralde, J. M. and Arca, J. M. (eds.), *Informe de Situación de la Economía Social Vasca*, Donostia-San Sebastián, Instituto de Derecho Cooperativo y Economía Social GEZKI (UPV/EHU), 2011.

Spear, R., "Formes coopératives hybrides," *RECMA, Revue internationale de l'économie sociale*, No. 320, 2011, pp. 26-42.

Statistics Canada, *Symposium 2005: Methodological Challenges for Future Information Needs*, Catalogue No. 11-522-XIE, Ottawa, 2005.

——, *Satellite Account of Non-Profit Institutions and Volunteering*, Catalogue No. 13-015-X, Ottawa, 2007a.

——, *North American Industry Classification System (NAICS) 2007*, Catalogue No. 12-501-XIE, Ottawa, 2007b.

——, *A Brief Guide to the Business Register*, Ottawa, Business Register Division, 2010.

——, *North American Industry Classification System (NAICS) 2007*, Catalogue No. 12-501-X, Ottawa, 2012.

Stiglitz, J., Sen, A. and Fitoussi, J.-P., *Report by the Commission on the Measurement of Economic Performance and Social Progress*, Paris, Commission on the Measurement of Economic Performance and Social Progress 2009.

United Nations, *Handbook on Non-Profit Institutions in the System of National Accounts*, Department of Economic and Social Affairs, Statistics Division, Series F, No. 91, New York, 2003.

——, *International Standard Industrial Classification of All Economic Activities (ISIC)*. Department of Economic and Social Affairs, Statistical Papers, Series M No. 5, Rev. 4, New York, 2008.

Organizing the Field of the Social Economy

The Social Economy and Its Classification within Systems of National Accounts

Édith Archambault

Professor emeritus, Centre d'économie de la Sorbonne, France

For it is not a question of linking consequences, but of grouping and isolating, of analysing, of matching and pigeon-holing concrete contents; there is nothing more tentative, nothing more empirical (superficially, at least) than the process of establishing an order among things; nothing that demands a sharper eye or a surer, better-articulated language; nothing that more insistently requires that one allow oneself to be carried along by the proliferation of qualities and forms. And yet an eye not consciously prepared might well group together certain similar figures and distinguish between others on the basis of such and such a difference: in fact, there is no similitude and no distinction, even for the wholly untrained perception, that is not the result of a precise operation and of the application of a preliminary criterion.

Michel Foucault, *Les mots et les choses*, 1966, pp. xxi

Introduction

Standard classifications were developed to simplify the complex world of the corporate sector, to analyze the international exchange of goods and services, and to enable cross-country comparisons of production sectors and products. They date back to the beginning of international exchange and are much older than the first system of national accounts (Vanoli, 2002). From a mathematical point of view, classifications are embedded partitions. They introduce discontinuities in a continuous reality (e.g., from the smallest to the largest producer) under the following hypothesis:

- There is much more similarity between items within a category than between those of two neighboring categories.
- Borderline and ambiguous cases are classified within one category, and only one, in the same way anytime and anywhere.

- Classifications can be aggregated or broken up, like Russian dolls.
- The standard classifications are revised periodically to reflect technological, institutional or organizational changes. These changes introduce a discontinuity in time series statistics. This is a great complication for historians or analysts of long-term trends, as they have to link successive time series. However, classifications need not be changed too often, because each revision breaks with the time series statistics serving as the basis of economic and social policy evaluations and forecasts.

This means that standard classifications are artefacts that, following their initial use, represent corporate producers more than unincorporated enterprises owned by households. They also serve to analyze goods more precisely than services. With regard to the non-market production of collective or divisible services, they also reflect the government more than non-profit institutions.

Social economy entities, being either market or non-market producers, must be included in the standard classifications, and they are – only not completely. In this chapter, I examine where in the national accounts these social economy entities are classified, and what are the advantages and drawbacks of these classifications. In addition, going beyond this standardization, I examine the role of the social economy in creating or repairing social ties, including the impact of these on society as a whole.

1. Advantages and drawbacks of standard classifications

This section examines the standard classifications applied to social economy units in the System of National Accounts (SNA 1993 and SNA 2008). For this, it refers to the national accounting framework and its classifications, despite the periodical resurgence of possibly unwarranted criticism of that framework.[1] National accounting cannot do more than what it was designed for. However, it has two valuable and irreplaceable qualities: it is unifying and empowering. It is unifying because it makes comparable different human activities by their purpose or their location, and because it allows gauging them across common quantitative scales. And it is empowering because national accounting, as a kind of universally accepted grammar, allows to formalize, estimate and elucidate complex realities or poorly understood interdependencies. Refined and customized by decades of use, national accounting can now integrate complementary and qualitative data for humanizing the monetary core

---

[1]   See the contribution by Mertens and Marée in this book.

framework (Archambault and Kaminski, 2009). This openness relies in particular on the satellite accounts (Stiglitz, Sen and Fitoussi, 2009).

1.1. Where are social economy units among the institutional sectors in national accounts?

A main mode of classification in national accounts is the distribution of economic actors among institutional sectors. This classification relies firstly on the main economic function of the unit (e.g., production, consumption, finance) and secondly on its main source of funding (e.g., sales, taxes, wages or other incomes). The guidelines of the international *System of National Accounts, 1993* (hereinafter SNA 1993) and the more recent *System of National Accounts, 2008* (hereinafter SNA 2008) categorize social economy units among the institutional sectors[2] of national accounts according to the following scheme. This also applies to their application to the *European System of Accounts* 1995 and 2010 (hereinafter ESA 1995 and ESA 2010).

Cooperatives[3] and non-profit institutions[4] (NPIs) whose income comes mainly (more than 50% in ESA 1995) from the sale of goods or services produced at market prices are classified as *non-financial corporations* or *financial corporations* according to the kind of their product." These social economy organizations must "continue to be treated as market producers as long as their fees are determined mainly by their costs of production and are high enough to have a significant impact on demand." The *non-financial corporations* sector also includes NPIs that are funded by or serve the interests of businesses, such as chambers of commerce or trade associations (SNA 1993, 4.58, 4.59).

Insurance mutual societies are classified as *financial corporations* providing life, accident, sickness and other forms of insurance to institutional units or groups of units within the sub-sector *insurance corporations and pension funds* (SNA 1993, 4.97, 4.98).

---

[2] According to SNA 1993, 2.19-20, "Institutional units are capable of owning assets and incurring liability on their own behalf. They are centers of legal responsibility and of decision-making for all aspects of economic life. These institutional sectors are grouped together to form institutional sectors on the basis of their principal function and main resources."

[3] According to SNA 2008, 4.41, "Cooperatives are set up by producers to organize the marketing of their collective output. The profits of such cooperatives are distributed in accordance with their agreed rules and not necessarily in proportion to shares held, but effectively, they operate like corporations."

[4] According to the SNA 2008, 4.83, "Non-profit institutions are legal or social entities created for the purpose of producing goods and services, but whose status does not permit them to be a source of income, profit or other financial gain for the units that establish, control or finance them."

NPIs providing goods or services free or at prices that are not economically significant to individual households are classified to *general government* if they are mainly (over 50% in ESA 1995) financed and controlled by the government sector. These non-market NPIs are further allocated to the sub-sector *state government* if they are controlled and mainly financed by the central government or public agencies, and to the *local government* sub-sector if they are controlled and mainly financed by local government units (SNA 1993, 4.62-63). In SNA 2008, "[t]he determining factor is whether the unit is part of, or controlled by, government," whereby the criterion concerning the origin of the NPI's sources of income disappeared (SNA 2008, 4.25). As a result of this major change from SNA 1993 to 2008, less NPIs are now classified under *general government*, and the *non-profit institution serving households* sector (hereinafter NPISH) is enlarged, as it now includes independent NPIs mainly funded by the government sector.

Small NPIs run by volunteers without a paid staff are allocated to the *households* sector and their current expenses are considered as final consumption, while in the case of NPIs allocated to other sectors, such expenses are considered as intermediate consumption (ESA 1995, 2.88).

Other non-market NPIs – that is, NPIs with at least one employee funded mainly by membership dues, donations or other earned income, and/or mainly funded but not controlled by central or local government units – are classified as NPISHs. This sector is composed on the one hand of member-serving organizations, such as professional or learned societies, political parties, labor unions, consumer associations, churches and religious societies, or social, cultural, recreational or sports clubs. On the other hand, another type of NPISH is composed of organizations created to serve a philanthropic or public purpose rather than the interests of their members. The resources of these charities or relief or aid agencies are individual or corporate giving and public or international funding (SNA 1993, 4.65-67). NPISH is therefore a residual sector and its accounts are often not filled by countries that lack basic statistical information on such institutions; in these cases, the operations that would normally qualify as NPISH units are either allocated to households or simply overlooked.

In national accounts, the social economy is thus subdivided into five institutional sectors, as summarized in Table 1.

Table 1. *The social economy in national accounts*

Social economy organizations	Institutional sector of SNA 1993
Non-financial cooperatives	Non-financial corporations, S11
Market non-profit institutions	Non-financial corporations, S11
Cooperatives and mutual banks, savings and loans and other social economy financial organizations	Financial corporations, S12

94

Insurance and health mutual societies	Financial corporations, S12
Non-market NPIs (mainly funded and) controlled by government units	General government, S13
NPIs with no employee	Households, S14
NPIs not elsewhere classified	Non-profit institutions serving households (NPISH), S15

Source: Archambault and Kaminski, 2009.

The advantage of the SNA classification is that all producers and consumers are grouped within one, and only one, institutional sector. All economic transactions are described in a sequence of current and accumulation accounts and balance sheets. The balancing items of these accounts are important aggregates such as value added, operating surplus, disposable income, savings and net worth.

The first obvious drawback of this classification is that it cannot capture the economic weight of the social economy, given its fragmentation across the institutional sectors. SNA 2008 affords a considerable improvement for NPIs in that it recommends that both financial and non-financial corporations be disaggregated, as in general government, to show NPIs as separate sub-sectors, so as to facilitate the derivation of a satellite account for NPIs (SNA 2008, 4.35). Another more general drawback is that economic exchange is privileged over other functions of social economy units. Thus, purely social ties or political influence are out of the scope, although they could be reintroduced through specific indicators of a social economy satellite account.

To overcome these disadvantages and facilitate the building of a NPI satellite account, SNA 2008 states:

> Like the 1993 SNA, the 2008 SNA assigns non-profit institutions (NPIs) to different institutional sectors, regardless of motivation, tax status, type of employees or the activity they are engaged in. Recognizing the increasing interest in considering the full set of NPIs as evidence of 'civil society,' the 2008 SNA recommends that NPIs within the corporate and government sectors be identified in distinct subsectors so that supplementary tables summarizing all NPI activities can be separately derived (SNA 2008, 4.35 and Annex 3.17).[5]

1.2. What do social economy units produce?

In the analysis of the economic functions of corporations and unincorporated enterprises, two international classifications are in use: the

---

[5] There remains the issue of distinguishing what part of civil society is considered as participating in the social economy. For this discussion, see Bouchard, Cruz Filho and St-Denis, 2011.

International Standard Industrial Classification of All Economic Activities (ISIC) and the Central Product Classification (CPC). Although a product is generally matched with an activity, a one-to-one correspondence between activities and products does not always exist, as certain activities produce more than one product (joint products) and as a product may be produced by using different techniques of production (SNA 2008, chapter 5). However, CPC is more detailed, offering five levels instead of the four offered by ISIC.

1.2.1. The International Standard Industrial Classification of All Economic Activities (ISIC)

In ISIC, corporations and other enterprises including general government entities, NPISHs and other social economy enterprises are allocated to an industry sector based on their principal activity.[6] An industry thereby gathers enterprises engaged in the same activity at the lowest level of the classification and in similar activities at most aggregated levels. The industry classifications of ISIC (and its adaptations to regional areas, such as NAICS for North America and NACE for Europe) are determined according to three criteria, by decreasing importance:

1. The physical composition and the stage of fabrication of the good or service
2. The use of the good or service
3. The inputs, processes and techniques of production

Of course, criteria 1 and 3 refer more to the production of goods and criterion 2 more to services. ISIC and its twin, the CPC, are frequently revised because they become quickly obsolete due to technological progress. For example, many IT products or e-trade and online services were added to the latest version of ISIC, ISIC Rev. 4, the most aggregated level of which is presented in Table 2.

Table 2. *International Standard Industrial Classification of All Economic Activities, Rev. 4*

A – Agriculture, forestry and fishing
B – Mining and quarrying
C – Manufacturing
D – Electricity, gas, steam and air conditioning supply
E – Water supply; sewerage, waste management and remediation activities

---

[6]   The principal activity may be determined on the basis of the highest value added or by default the highest turnover or number of employees.

F– Construction
G – Wholesale and retail trade; repair of motor vehicles and motorcycles
H – Transportation and storage
I – Accommodation and food service activities
J – Information and communication
K – Financial and insurance activities
L – Real estate activities
M – Professional, scientific and technical activities
N – Administrative and support service activities
O – Public administration and defense; compulsory social security
P – Education
Q – Human health and social work activities
R – Arts, entertainment and recreation
S – Other service activities
T – Activities of households as employers; undifferentiated goods- and services-producing activities of households for own use
U – Activities of extraterritorial organizations and bodies

http://unstats.un.org/unsd/cr/registry/regcst.asp?Cl=27&Lg=
Source: SNA 2008, Annex 1.

The social economy does not operate in all of these industries, and the type of or quantity of sectors it engages in differs from one country to the next. In developed countries, social economy units are nearly non-existent in the B to E, the O and the S to U industries. They are, however, numerous in the agriculture forestry and fishing, financial and insurance, education, human health and social work as well as the arts, entertainment and recreation industries.

The main advantage of ISIC is that it allows for a cross-country comparison of the social economy. This is because the detailed explanatory notes included in every classification are a guarantee that, roughly speaking,[7] all countries allocate the same activities to the same industries. ISIC also allows comparing social economy entities with other enterprises and to thereby calculate their "market share." Table 3, issued by the French National Institute for Statistics and Economic Studies, gives an example of such an application for France.

---

[7]   Any bilateral or multilateral comparison must take into consideration the habits of statisticians of each country. In that context, Flacher and Pelletan (2007) emphasize that differences in practices between Europe and America do exist, yet also that these differences are getting smaller due to convergences in the classification of activities.

Table 3. *Employment in the social economy as percentage
of total employment by industries, 2010*

Industries	% of total employment
A Agriculture, forestry and fishing	4.4%
B to F Manufacturing industries + construction	1.1%
of which Manufacture of food, beverages and tobacco products	4.7%
G to I Trade, transportation and accommodation	1.8%
of which Trade	1.9%
K Financial and insurance activities	30.2%
J, L to N Information, real estate, professional and support activities	4.2%
P Education	20.0%
Q Human health and social work activities	18.6%
of which human health	11.4%
of which social work	62.4%
R Arts, entertainment and recreation	42.9%
TOTAL Social economy	10.3%

Source: INSEE-CLAP Tableaux harmonisés de l'économie sociale 2010.

The table shows that in France, the social economy produces mainly services, which is consistent with findings for other countries. The French production of goods by cooperatives is concentrated in agriculture and food manufacturing. Cooperatives and mutuals are very active in the financial and insurance industry, while NPIs comprise a significant share of the service industries associated to the welfare state, such as education, human health and social work activities. Arts, entertainment and recreation services are mainly run by associations.

Returning again to ISIC, the main input in the production of services is labor. In these labor-intensive industries, paid employment is clearly recorded. However, volunteer work, a major input for associations, foundations and, to a lesser degree, for mutuals and cooperatives, is overlooked. As volunteering is not proportional to paid employment among the diverse activities, the structure of industries of NPIs, and therefore of the social economy, is skewed. This is a first disadvantage of the ISIC classification scheme.

Another disadvantage is that the ISIC classification by main economic activity is little adapted to the NPIs, who advocate or defend a cause more than provide goods or services. Indeed, ISIC was designed to provide a detailed representation of the market economy, with a greater focus on the production of goods than that of services. While the scheme is periodically revised to incorporate goods and services that were newly introduced due to technological progress, it remains poorly adapted to the non-market production of the government sector and the bulk of NPI services. The more recent classifications, ISIC Rev. 4 and CPC Ver. 2, are advanced compared to their predecessors, although they remain marked by their origin.

1.2.2. The Central Product Classification (CPC Ver. 2)

As mentioned above, CPC is nearly identical with ISIC but is more detailed at the lowest level of the classification. Particularly, CPC Ver. 2 provides a fairly accurate breakdown of community, social and personal services corresponding to social work activities in the 932 to 935 codes, as presented in Table 4.

Table 4. *Breakdown of community, social and personal services*

Hierarchy Section: 9 – Community, social and personal services Division: 93 – Human health and social care services
Breakdown: This Division is divided into the following Groups: 931 – Human health services 932 – Residential care services for the elderly and disabled 933 – Other social services with accommodation 934 – Social services without accommodation for the elderly and disabled 935 – Other social services without accommodation

Source: SNA 2008, Annex 1.

Another decisive improvement for the classification of NPIs would consist of a very fine breakdown of the services provided by *other membership organizations not elsewhere classified*. In most countries, the use made of this residual position tends to be too general to be of value, reflecting the unsuitableness of the classification, difficulties in determining the main activity of an organization performing multiple activities, if not sheer laziness on the part of the coder. The breakdown of code 9599 down to its five-digit subclasses is presented in Table 5.

Table 5. *Breakdown of the services furnished*
by other membership organizations

Hierarchy Section: 9 – Community, social and personal services Division: 95 – Services of membership organizations Group: 959 – Services furnished by other membership organizations Class: 9599 – Services furnished by other membership organizations n.e.c.
Breakdown: 95991 – Services furnished by human rights organizations 95992 – Services furnished by environmental advocacy groups 95993 – Other special group advocacy services

95994 – Other civic betterment and community facility support services
95995 – Services provided by youth associations
95996 – Grant-giving services
95997 – Cultural and recreational associations (other than sports or games)
95998 – Other civic and social organizations
95999 – Other services provided by membership organizations

Source: SNA 2008, Annex 1.

1.2.3. The International Classification of Non-profit Organizations

With the aim to fill the gaps of ISIC Rev. 3,[8] the international research team of the Johns Hopkins Comparative Non-profit Sector Project tested and adopted the International Classification of Non-profit Organizations (ICNPO). This ad hoc classification, which contains 12 groups and 30 subgroups, can be embedded into ISIC and CPC.

This coherence has a double advantage. Depending on the needs and specificities of each country, the subgroups can be given titles that make sense for that country. For example, the addition of subgroups such as *popular education* or *social tourism* within *culture and leisure activities* makes sense in Nordic and French-speaking countries.

It should be noted that ICNPO is relevant only for NPIs. Another advantage of its integration in ISIC when analyzing the entire social economy, ISIC allows to report specific activities of cooperatives and mutual companies, such as food-processing industries, trade, insurance services and financial activities.

Table 6. *International Classification of Non-profit Organizations (ICNPO)*

1 Culture and recreation
11 Culture (media and communication, arts, performing arts; museums, learned societies)
12 Sports
13 Recreation (recreation, social tourism, service clubs)
2 Education and research
21 Primary, secondary and higher education.
22 Other education (training and adult education, alumni, parent-teacher NPIs)
3 Health
31 Hospitals and rehabilitation; nursing homes
32 Other health services (crisis intervention, sanitary education, emergency, self help)

---

[8] ISIC Rev. 3 was in use before ISIC Rev. 4. The latter benefitted from the critiques made of the former.

4 Social services
41 Residential homes (for the disabled, elderly, homeless, etc.)
42 Social services without accommodation (income support and material assistance, day care, child and family welfare, home services; emergency and relief charities)
5 Environment
(pollution control, natural resources conservation, animal protection)
6 Development and housing
61 Economic, social and community development
62 Building or rehabilitation; housing of students, workers [...] and assistance
63 Employment and on-the-job training, vocational rehabilitation
7 Law, advocacy and politics
71 Civic and advocacy NPIs
72 Law and legal services (crime prevention and rehabilitation, victim support, consumer protection)
73 Political organizations
8 Philanthropic intermediaries and voluntarism promotion
9 International activities (exchange programs, development assistance and relief)
10 Religion (religious congregations and associations)
11 Business and professional associations and labor unions
12 Not elsewhere classified (n.e.c.)

Source: Salamon, Sokolowski *et al.*, 2004.

ICNPO, which is based mainly on the nature of the services provided, showed itself well adapted to its objective, as testified by the fact that the residual category 12 was void in most of the 36 countries that participated in phase 2 of the above-mentioned project. Indeed, the quality of a nomenclature can be estimated by the degree to which its residual category is *not* used. Given its success, ICNPO was adopted, with minor modifications, by the *Handbook on Non-Profit Institutions in the System of National Accounts*, published by the United Nations in 2003.[9]

1.3. What kind of functions do social economy entities fulfill?

In the SNA, functional classifications are proposed to identify the purpose or the objectives of non-market producers (central and local governments and NPISHs), who comprise a majority of the social economy entities. These classifications suggest that the government and NPISHs respond to the needs of the population through the provision of collective goods or private goods with positive externalities.

---

[9] Since there are no categories in this classification for activities that are typically organized by cooperatives and mutual societies, namely market-oriented activities, this classification needs to be extended or modified when seeking to cover the whole of the social economy. ISIC is a good basis for proceeding with such an extension. For an example of such a classification, see Bouchard *et al.*, 2008.

For NPISHs, the Classification of the Purposes of Non-Profit Institutions Serving Households (COPNI) is useful in describing the various outlays of NPISHs according to the social need fulfilled. However, hardly any countries collect such detailed information on these entities, nor is COPNI really in use. This is a pity, because the COPNI of SNA 2008 is more detailed than its predecessor, SNA 1993, and much closer to ICNPO.

Table 7. *Classification of the Purposes of Non-Profit Institutions Serving Households (SNA 2008)*

01 – Housing
     – Housing
02 – Health
     02.1 – Medical products, appliances and equipment
     02.2 – Outpatient services
     02.3 – Hospital services
     02.4 – Public health services
     02.5 – R&D Health
     02.6 – Other health services
03 – Recreation and culture
     03.1 – Recreational and sporting services
     03.2 – Cultural services
04 – Education
     04.1 – Pre-primary and primary education
     04.2 – Secondary education
     04.3 – Post-secondary non-tertiary education
     04.4 – Tertiary education
     04.5 – Education not definable by level
     04.6 – R&D Education
     04.7 – Other educational services
05 – Social protection
     05.1 – Social protection services
     05.2 – R&D Social protection
06 – Religion
     06.0 – Religion
07 – Political parties, labor and professional organizations
     07.1 – Services of political parties
     07.2 – Services of labor organizations
     07.3 – Services of professional organizations
08 – Environmental protection
     08.1 – Environmental protection services
     08.2 – R&D Environmental protection
09 – Services not elsewhere classified (n.e.c.)
     09.1 – Services n.e.c.
     09.2 – R&D Services n.e.c.

http://unstats.un.org/unsd/cr/registry/regcst.asp?Cl=6&Lg=1
Source: SNA 2008, Annex 1.

A potential advantage of COPNI is its inclusion of R&D services and of advocacy as purposes. The classification is also well suited to the comparison of the role of government (through COFOG, the classification of total outlays of government by function)[10] with that of NPISHs in providing collective or quasi-collective goods. It is especially significant when government units cannot or do not desire to provide these collective or quasi-collective goods and services. The comparison of COPNI with the Classification of Individual Consumption According to Purpose (COICOP) is also feasible and significant in terms of the welfare and living conditions of households.

Yet, COPNI also has disadvantages. For example, a purpose is less quantifiable than a product, even if it is a service product, and it is less material than a good. A purpose must first be declared, and the declared purpose may differ from the final product. Finally, how well can COPNI capture multi-purpose NPIs? In companies pursuing multiple activities, the company's main activity is the activity consuming the greatest portion of the company's value added or turnover (in default employment). Yet this approach does not work for multi-purpose NPIs, and would result in an inflation of the residual category 09 (*services not elsewhere classified*).

2. Beyond the standard classifications

2.1. Measuring the social ties that social economy units create

The standard classifications are designed to serve national accounts, namely by describing all the transactions that contribute directly or indirectly (through intermediate production) to the gross domestic product (GDP) of a country. Given this focus on GDP, these classifications invariably overlook the social ties that market transactions and non-market cash or in-kind transfers afford. Yet for social economy entities, these social ties come first because they associate persons and not funds, contrary to stock companies and corporations. Social economy units are main building blocks of the social capital, as demonstrated by Putnam (Putnam, 2000). These social ties are rarely measured, but when they are, specific classifications are needed in order to accurately reflect the type of governance and ownership of the social economy unit.

2.1.1. Measuring the membership of the social economy

The social economy exhibits three main types of memberships in which members have an additional function. One, in production cooperatives, the members are generally employees at the same time. Two, in

---

[10] United Nations Statistics Division, *Classification of the Functions of Government*, on-line: http://unstats.un.org/unsd/cr/registry/regcst.asp?Cl=4&Top=2&Lg=1.

most cooperatives and mutual insurance companies (consumer cooperatives, cooperative or mutual banks, mutual damage, health or life insurances) and in NPIs working in their members' interest, the members are
also clients. And three, in cooperatives that represent or associate independent workers or unincorporated organizations (farmers, craftsmen or
merchants), the members are either clients for intermediate products or
sellers of their own product.

Within these three types of membership, the double function held by
the members provide these with greater opportunities to generate social
ties. For example, members here are invited to attend more meetings,
including the annual general meetings, where their organization's governing body is elected and where its activities, output and management are
reported on. Of course, the social ties are weaker in large social economy
organizations than in small ones.

The social economy also exhibits a fourth type of membership in
which members can also be beneficiaries, yet where these two functions
are more explicitly disassociated from one another. This applies mainly
to associations working for the public interest or the common good.[11]
Here, the members do not derive a direct benefit from belonging to the
NPI, although they do have the opportunity to build social ties by attend-
ing meetings, as in member-oriented NPIs. In addition, they can increase
this opportunity to create social ties by serving as volunteers for their
organization.

This distinction between these four categories of members could be
the beginning of a classification of members. At the international level,
NPIs generally have greater knowledge on their membership than cooperatives and mutuals. Still, in Anglo-Saxon countries, NPIs' data on and
knowledge of their membership base is common and provides them with
long time series, although these data are rarely comparable across countries. In the European Union, the situation is somewhat different owing to
two surveys: the Survey on Income and Living Conditions of Households
(SILC) and, less reliable, the European Values Study. These surveys break
down membership according to members' socio-demographic characteristics and to the industry sector of their NPI. Unfortunately, the allocation to the industry sector in these surveys is inadequate, being extremely
broad as well as incomplete or obsolete.[12]

---

[11] Foundations have no members.

[12] For instance, in SILC 2008: health or social services organizations, charities; sports;
culture; recreation; advocacy and lobbying associations; elderly clubs; unions and professional associations (SILC 2008).

2.1.2. Measuring and classifying volunteer work

Volunteer work is not yet within the scope of national accounts. As unpaid household activities performed mainly by women (Stiglitz *et al.*, 2009), which still falls outside of the borderline (or perimeter) of production in use. Nevertheless, SNA 2008 does offer a somewhat wider definition of production than its predecessor, SNA 1993:

> Economic production may be defined as an activity carried out under the control and responsibility of an institutional unit that uses inputs of labour, capital and goods or services [...] Activities that are not productive in an economic sense include basic human activities such as eating, drinking, sleeping, taking exercise etc., that it is impossible for one person to employ another person to perform instead (SNA 2008, 6.24-25).

At present, the production of volunteer services is indeed carried out under the control and responsibility of social economy units. These organizations use inputs of labor, capital and intermediate goods or services. In addition, volunteer work answers the criterion of the third party, because we can generally substitute it with some paid work (Hawrylyshyn, 1977). Volunteering is thus situated, as household work, somewhere between the wide and narrow boundaries of production.

In addition, none of the following justifications of SNA 2008 for excluding the own-account production of services within households from the narrow boundary of production is valid for the volunteer work of social economy organizations:

> the relative isolation and independence of these activities from markets, the extreme difficulty of making economically meaningful estimates of their values, the adverse effects it would have on the usefulness of the accounts for policy purposes and the analysis of markets and market disequilibria (SNA 2008, 6.29-30).

Indeed, the activities of volunteers are independent neither from the market of goods and services nor from the labor market. It is not impossible to attribute an economic value to the time of volunteer work, because this work, being socially organized, is more easily comparable to paid work than unpaid household work; its monetary valuation is thus less arbitrary (Archambault and Prouteau, 2009). Finally, because the weight of volunteer time is much smaller than unpaid household time, it does not modify the labor market equilibrium. In addition, economic and social policy obviously influences social economy organizations and volunteers; sometimes, social policies are even discussed with social economy organizations and tested through or by them.

Overall, there is a trend to measure volunteer work taking place in mutual societies and NPIs. Volunteer work is deemed to contribute significantly to the "added value" of these organizations, even if in more professionalized ones, volunteering is limited to serving as a board member or other type of elected member. Volunteers are frequently members of the organization, and in most countries, one out of two members engage in some kind of volunteer task either regularly or occasionally (SILC 2008). Yet, membership is not a prerequisite to volunteering; for example, many people contribute to an organization only occasionally, such as for special events, or on a regular basis without being a member of the organization.

As existing surveys on volunteering are neither regular nor comparable, the International Labor Organization (ILO) calls for, in its *Manual on the measurement of volunteer work*, the adoption of common methodology and system of classification (ILO, 2011). In the following, we will present and examine the definition, delimitation and classification of volunteer work presented in this Manual.

The definition of volunteer work is as follows:

Unpaid non-compulsory work; that is, time individuals give without pay to activities performed either through an organization or directly for others outside their own household[13] (ILO, 2011, 3.5).

This international definition is more comprehensive than that of most existing surveys, which generally concern only organized volunteer work. It also includes direct assistance given to persons outside the volunteer's household and family. However, only organized type of volunteer work is of interest for the purposes of this article. Volunteer work can be performed for a range of causes, including for people, animals, the environment and the wider community. It provides divisible as well as collective goods and services. Volunteering can also benefit organizations outside of the social economy, in particular government agencies, local communities and even private companies.

Any cases bordering on other activities must, as is a principle in all classifications, be either included or excluded from the volunteering category. The definition specifies, first, that volunteer work is not compulsory by law or imposed by physical force. As such, it excludes, for example, community work done instead of serving a prison sentence. However, volunteer work performed under strong social pressure cannot be disqualified as volunteer work.

---

[13] UN, 2003, 4.45 gives the following definition: "work without monetary pay or legal obligation provided for persons living outside the volunteer's household".

Secondly, the border between volunteering and leisure is based on the above quoted third-party criterion (Hawrylyshyn, 1977). This means that giving free tennis lessons, for example, is volunteering because the volunteer can be replaced by a paid coach. By contrast, playing tennis is leisure because nobody would pay someone to play instead of him/her. However, the third-party criterion works less well for more militant activities or those with extremely strong emotional attachments. For example, I cannot pay someone to demonstrate in the street to defend a cause, or to visit an ill person in a hospital instead of me. Yet, these activities are considered to be traditional volunteer tasks in most countries.

Thirdly, the borderline between volunteering and paid work is easier to distinguish given the existence of a contract and a wage with paid employment. One exception may be freelance work, which often brings no earnings in the beginning. Further gray zones between a wage earner and a volunteer also complicate the categorization to any one group. Among these are the too generous reimbursement of expenses, fringe benefits and civic volunteer services of youth indemnified below the minimum wage.

The borderline to training or education has to be clarified as well. This should obviously exclude the unpaid time spent to study, as students cannot pay somebody to study for them. However, unpaid internships or student volunteer work in social economy or other organizations falls within the scope of volunteering if carried out voluntarily, but outside if it is compulsory in the sense of being an integral part of the academic requirements for that student.

To fill these gaps, the above-mentioned *Manual on the measurement of volunteer work* recommends that statistical offices add a brief volunteer supplement to their respective Labor Force Survey, which is an internationally standardized data collection program. In this way, volunteer work and paid work can be observed in the same industrial and occupational classification framework.

Since ISIC Rev. 4, together with its national and multi-national counterparts, is the main classification used in the Labor Force Surveys, the Manual recommends it for identifying the industry of a given volunteer activity, particularly since ISIC Rev. 4 now incorporates much of the detail formerly available only in ICNPO (see section 1.2 of this article). Yet, ICNPO can be used as well, and the Manual provides a detailed table of linkages, or "crosswalks," between the two classifications.

The International Standard Classification of Occupations (ISCO-O8), or its national or regional equivalents, is recommended by the ILO Manual to classify volunteer work activities. This classification contains four levels, the first two of which are sufficient for classifying volunteer work. The following Table 8 gives examples of types of volunteer work, referred to as "volunteer occupations," associated with major groups of ISCO. ISCO-08 allows to achieve a reasonable degree of comparability of the data collected in different countries despite the various traditions of volunteering.

Table 8. *Examples of volunteer occupations associated with ISCO-08 major groups*

ISCO major group	Examples of volunteer occupations
1. Legislators, senior officials and managers	Lead or manage a non-profit organization, association, union, or similar organization. Serve on a board of directors or management committee of a social economy organization Policy and research managers
2. Professionals	Develop emergency plans for a community Provide pro bono legal or dispute resolution services Manage a programme or organization designed to collect and analyze data for public information Provide professional social work and counseling services
3. Technicians and associate professionals	Provide emergency medical care Take the lead in planning, managing, or organizing an event Coach, referee, judge, or supervise a sports team Teaching, training, or tutoring
4. Clerks	Interview other people for the purpose of recording information to be used for research Provide clerical services, filing and copying Help to provide technical assistance at a sporting or recreational event
5. Service workers and shop and market sales workers	Prepare or serve meals for a soup kitchen Contact people to advance a cause by going door-to-door Help and entertain children in a summer camp Sell in a charity shop
6. Skilled agricultural workers	Make improvements to the public green areas of a community, by planting trees and other nursery stock Care birds after an oil spill
7. Craft and related trades workers	Construction, renovation and repairs of dwellings and other structures in a cooperative or a community development non-profit Bicycle repair and maintenance in a sports club

8. Plant and machine operators and assemblers	Drive children to a sporting or recreational event Drive a film projector in an elderly club, a cine-club.
9. Elementary occupations	Collect trash, garbage and sort recycling materials Help to clean up after a sporting or recreational event for public entertainment Do odd jobs for a non-profit organization

Source: Adapted from ILO 2011, Table 5.1.

What are the advantages and the drawbacks of ISCO? The main advantage is comparability, followed by the ability to cross-reference commonly performed volunteer occupations between ISCO and its national or multi-national equivalents. Such matrices are key instruments for matching the verbatim responses of survey participants with the appropriate codes of the classification. The third main advantage of ISCO is that it facilitates the task of assigning a monetary value to volunteer work, namely by proposing the use of the average wage of the performed occupation.

Yet, there are also drawbacks of ISCO. First, some types of volunteer work are difficult to classify. For example, in organizations that have no paid staff at all, volunteers often fulfill a wide range of tasks, from leading and managing their organizations to cleaning up after the meetings. This is often the case with grassroots organizations, which exist in all countries. Other examples include, as mentioned above, occupations that are more militant or emotionally attached.

The second drawback concerns the use of the average wage of an occupation as the shadow wage of a volunteer. The problem with this approach is that it presumes that the productivity of a volunteer and a paid employee is the same. Yet that premise is obviously false, with the possible exception of volunteering in the same occupation as one's real job or profession. Indeed volunteers are generally less qualified than employees and spend more time socializing and creating social ties with the beneficiaries of the organization.

2.2. Measuring the impact of social economy entities involves creating ad hoc indicators

Social economy entities are increasingly asked to account for their economic, social, environmental and societal impact to their stakeholders, such as owners, employees, volunteers, central and local public funders and donors. For this, they need tools for evaluating their performance other than the standard yardstick, the rate of profit – and they find them by building, on an ad hoc basis, multi-dimensional indicators. Subsequently, these indicators can be used in reference classifications common to the organizations working in the same field, the same area or competing for the same bidding offer.

Such classifications are ideally built in partnership between the organizations and their partners. The integration of ad hoc indicators into standard classifications is not necessary, as the aim of the indicators is benchmarking with organizations working in the same industry or for the same public. Of course, these ad hoc classifications are more normative than positive. Still, this is a lesser problem once the normativity is known and taken into account by the users of the classification. A good indicator has to be relevant to the purpose of the social economy organization; it must be simple to understand by the stakeholders, including the volunteers; and it must be calculated in the same way over a span of time long enough to present a reliable evolution.

The following Table 9 shows how these indicators could be articulated to measure the social utility (or public interest) of social economy organizations (Gadrey, 2003) according to five multi-dimensional main themes, divided into two further sublevels, the finest one consisting of one or more indicators.

Table 9. *Classification of a multi-dimensional social utility*

Theme 1	Global criteria	Elementary criteria
Social utility with strong economic component	Created or saved economic wealth	*Lesser collective cost*
		Indirect reduction of costs
		Contribution to the rate of activity
	Territory	*Contribution to the economic dynamism*
		Liveliness of the community, the district

Theme 2	Global criteria	Elementary criteria
Equality, human development and sustainable development	Equality, development of "capabilities"	*Reduction of social inequalities*
		Actions towards disadvantaged public
		Insertion of the long term unemployed in the employment
		Professional equality men/ women
		Tiered pricing for the services
		Right in the housing
		Remedial courses for children in trouble
		Resumed self-confidence
	International solidarity, human development	*Actions for the development and struggle against poverty*
		Defense of human rights
	Sustainable development	*To improve the quality of the natural environment*
		To protect natural resources

Theme 3	Global criteria	Elementary criteria
	Social link	*Creation of social links*
		Mutual aid, local exchanges of knowledge
Social link and		*Positive impact of the social capital*
local democracy	Local democracy	*Participative dialogue, process of pluralistic decision making*
		Voicing of opinions of the citizens

Theme 4	Global criteria	Elementary criteria
	Innovation	*Discovery of emergent needs*
Contributions	Value of the	*Innovative ways of coping with unsatisfied needs*
in the social,	**"world" of the**	
economic and	creation	*Institutional innovations*
institutional		*Organizational innovations*
innovation		*Distinction of the internal and external innovations*

Theme 5	Global criteria	Elementary criteria
	Not for profit,	*Non-profit management*
	giving and	*Volunteer board*
Internal social	**volunteering**	*Voluntary action*
utility with	**More democratic**	*Rules of internal democracy and joint participation*
possible effects of	**and alternative**	
external contagion	**governance**	*Free membership: free entrance and free exit*
	Associative professionalism	*Volunteers professionalism*
		Cooperative internal training
		Social and wage acknowledgement
		Internal and external trainings

Source: Archambault, Accardo, Laouisset, 2010, adapted from Gadrey, 2003.

The partnerships between the central or local governments and the social economy organizations are institutional arrangements that are regulated by rules, formal and informal, and procedures. Several typologies for analyzing these institutional arrangements have been developed, the most prominent one being the one by Ostrom and Crawford. A summarized version of this typology is presented in Table 10 (Ostrom and Crawford, 2005; Elbers and Schulpen, 2013).

Table 10. *Classification of the rules and content of a partnership*

Type of rule	Content	Key questions
Boundary	Entry and exit	Which type of actors may participate? Who decides who is in and who is out? Which criteria are used for selection?
Scope	Outcomes	What are the outcomes to be achieved? What characteristics should outcomes have?
Position	Roles	What positions exist? What responsibilities are associated with these positions?
Choice	Actions	What are the rights and obligations of different actors?
Aggregation	Decision-Making	What is the level of actors' participation in decision-making? On which topics do they participate and in which decision making stage?
Information	Information exchange	What type of information do actors have to exchange? How frequently do actors have to exchange information?
Pay-off	Performance	How is performance defined and measured? What are the consequences of excellent or poor performance?

Source: Elbers and Schulpen, 2013, adapted from Ostrom and Crawford, 2005.

Conclusion

Classifications structure our vision of reality. As Giddens pointed out, structures are rules and resources at the same time (Giddens, 1984). The social economy is obliged to resort to standard classifications when undertaking cross-country comparisons. Yet it has to innovate and adapt existing classifications when seeking to analyze other fields of action, such as reporting on its specificity, values, volunteer work or its alternative way of governance. In all cases, however, one should not ask more of classifications than they can give.

I return, in conclusion, to the epigraph at the beginning of this article by Michel Foucault, which states that a classification is "an institution of an order among things." Every institution relies on the temporary agreement of its stakeholders (Desrosières and Thévenot, 2002). And those agreements, or rules, have a self-disciplinary effect (Foucault, 1975) whereby the artefact may eventually be taken as a self-evident truth.

The history of classifications (Guibert, Laganier and Volle, 1971; Desrosières, 2000; Desrosières and Thévenot, 2002) shows how classifications, as artefacts, shape our vision of the economic and social reality while also eclipsing a part of this reality, namely the social economy.[14] Overall, the study of classifications, including those of the past, the present and how they succeed and overlap each other over time, reveals the discretionary character of classifications. The three notable statisticians Guibert, Laganier and Volle, aptly expressed this realization as follows:

> The economist is not interested in, to use a metaphor, the glasses through which he sees the economy. Rather, he is interested in what he sees. To see the glasses we wear, we need to remove them first, which blurs our view. Also, discussions about classifications render the aggregates that had guaranteed solidity up to then as fragile, modifiable and rather doubtful. Outlines that were once clear become unpleasantly fuzzy [our translation] (Guibert, Laganier and Volle, 1971).

References

Archambault, E., *The Non-Profit Sector in France*, Manchester, Manchester University Press, 1997.

Archambault, E. and Kaminski, P., "La longue marche vers un compte satellite de l'économie sociale," *Annals of Public and Cooperative Economics*, 2009, Vol. 80, issue 2, pp. 225-246.

Archambault, E. and Prouteau, L., "Mesurer le bénévolat pour en améliorer la connaissance et satisfaire à une recommandation internationale," *RECMA, Revue internationale de l'économie sociale*, 2009, Vol. 314, pp. 84-104.

Archambault, E., Accardo, J. and Laouisset, B., *Rapport du groupe de travail "Connaissance des associations,"* No. 122, Paris, CNIS, 2010.

Barea, J. and Monzón, J. L., *Manuel pour l'établissement des comptes satellites des entreprises de l'économie sociale: coopératives et mutuelles*, Liège, CIRIEC, 2006.

Bouchard, M. J., Ferraton, C., Michaud, V., "First Steps of an Information System on the Social Economy : Qualifying the Organizations", *Estudios de Economía Aplicada*, vol. 26 no 1, April 2008, pp. 7-24.

Bouchard, M. J., Cruz Filho P. and St-Denis, M., *Un cadre conceptuel pour déterminer la population statistique de l'économie sociale au Québec*, Report for the Institut de la statistique du Québec. Montréal, UQAM, Canada Research Chair on the Social Economy and CRISES, No. R-2011-02, 2011.

Desrosières, A., *La politique des grands nombres: Histoire de la raison statistique*, Paris, La Découverte, 2000.

Desrosières A. and Thévenot, L., *Les catégories socio-professionnelles*, Paris, La Découverte, 2002.

European Commission, *European System of Accounts, ESA 1995*, Luxembourg, Eurostat, 1995.

---

[14]   See in this book the chapter signed by Artis, Bouchard and Rousselière.

European Commission, *European System of Accounts, ESA 2010*, Luxembourg, Eurostat, 2010.

Elbers, W. and Lau, S., "Corridors of Power: The Institutional Design of North-South NGO Partnerships," *Voluntas*, 2013, Vol. 24, pp. 48-67.

Flecher, D. and Pelletan, J., "Le concept d'industrie et sa mesure: origine, limites et perspectives. Une application à l'étude des mutations industrielles," *Economie et Statistique*, 2007, Vol. 405-406, pp. 13-46.

Foucault, M., *Surveiller et punir. Naissance de la prison*, Paris, Gallimard, 1975.

Foucault, M., *Les mots et les choses. Une archéologie des sciences humaines*, Paris, Gallimard, 1966.

Foucault, M., *The Order of Things*, 1966, New York, Random House, Vintage Books Editions, April 1994.

Gadrey, J., *L'utilité sociale des organisations de l'économie sociale et solidaire*, Summary report for DIES and MIRE, 2003.

Giddens, A., *The Constitution of Society, An Outline of the Theory of Structuration*, Berkeley and Los Angeles, University of California Press, 1984.

Guibert, B., Laganier, J. and Volle, M., "Essai sur les nomenclatures industrielles," *Économie et Statistique*, 1971, Vol. 20, pp. 21-36.

Hawrylyshyn, O., "Towards a Definition of Nonmarket Activities," *Review of Income and Wealth*, 1977, Vol. 23, pp. 79-86.

International Labor Office, *Manual on the Measurement of Volunteer Work*, Geneva, 2011.

INSEE, *Tableaux harmonisés de l'économie sociale*, http://www.INSEE.fr/fr/themes/detail.asp?ref_id=eco-sociale.

Kaminski, P., *Le compte des institutions sans but lucratif en France*, Paris, INSEE, Rapport de mission, 2005.

Ostrom, E. and Crawford, S., "Classifying Rules," in E. Ostrom (ed.), *Understanding Institutional Diversity*, Princeton, Princeton University Press, 2005.

Prouteau, L. and Wolff, F.-C., "Le travail bénévole; un essai de quantification et de valorisation," *Economie et Statistique*, 2004, Vol. 373, pp. 33-56.

Putnam, R., *Bowling Alone. The Collapse and Revival of American Community*, New York, Simon and Schuster, 2000.

Salamon, L. M. and Anheier, H. K., "In Search of the Nonprofit Sector II: The Problem of Classification, 1992," *Voluntas, Vol. 3, No 3*, pp. 267-309.

Salamon, L. M. and Anheier, H. K, "The International Classification of Nonprofit Organizations – ICNPO. Revision 1.0," *Johns Hopkins University Nonprofit Sector Project*, Working paper No. 19, 1996.

Salamon, L. M., and Sokolowski, W. S. (eds.)., *Global Civil Society. Dimensions of the Non-profit Sector*, Vol. 2, West Hartford, Kumarian Press, 2004.

Stiglitz, J., Sen, A. and Fitoussi, J.-P., *Mesure des performances économiques et sociales*, La Documentation Française, 2009.

United Nations Statistics Division, *Handbook of Non-Profit Institutions in the System of National Accounts*, Statistical Papers, series F, No. 91, 2003.

United Nations, European Commission, International Monetary Fund, OECD and World Bank, *System of National Accounts 1993*, New York, 1994.

United Nations, European Commission, International Monetary Fund, OECD and World Bank, *System of National Accounts 2008*, New York, 2009.

Vanoli, A., *Une histoire de la comptabilité nationale*, Paris, La Découverte, 2002.

International Comparisons

Lessons from the Measurement of the Nonprofit Component

Lester M. SALAMON

Professor, Johns Hopkins University, and director,
Johns Hopkins Center for Civil Society Studies, USA

S. Wojciech SOKOLOWSKI

Senior research associate for the Johns Hopkins
Center for Civil Society Studies, USA

Megan A. HADDOCK

International Research Projects Manager at
the Johns Hopkins Center for Civil Society Studies, USA

A global "associational revolution" (Salamon, 1994) has swept the globe in the last three decades, resulting in a massive upsurge of private, voluntary activity in virtually every corner of the globe. The product of new communications technologies, popular demands for high quality public services, dissatisfaction with the operations of both the market and the state in coping with the inter-related social and economic challenges of our day, the availability of international assistance, and, last but not least, the rapidly growing numbers of educated individuals eager to use their skills and knowledge to make a social impact, this "associational revolution" has focused new attention and energy on the broad range of institutions and activities that occupy the social space between, or beyond, the market and the state.

The emergence of this set of institutions has significantly changed how we think about the relationships among the market, the state, and society, and triggered a new momentum that has helped to usher in democratic governments to over half of the global population and empower previously excluded segments of society. Previously thought to be largely restricted to the developed West, the civil society sector has turned out to be a major economic force spread throughout the globe, providing a wide range of services from education and health to social assistance, and facilitating numerous forms of social activity, from culture and recreation, to issue advocacy, political and occupational interest representation, and the promotion of international cooperation.

The enhanced political prominence and legitimacy of this range of phenomena has led to several efforts to conceptualize, understand, and analyze it. However, the tremendous diversity of this set of institutions, and activities and the novelty of treating them as a coherent sector, coupled with deep-seated ideological preconceptions, has often led to confusion and debate about what the sector includes and how it is defined, prerequisites that are needed for a comparative study of the nonprofit sector.

Currently, the only internationally accepted official definition of the civil society sector institutions is found within the framework of the System of National Accounts (SNA), the internationally agreed standard set of recommendations on how to compile measures of economic activity in accordance with strict accounting conventions based on economic principles. The SNA identifies a distinct set of units called non-profit institutions (NPIs) defined by three key features: first, they are "organizations" i.e., economic units recognized, by law or custom, as separate entities that engage in economic transactions that thus have, or could have regular economic accounts; second, these entities are not allowed to distribute any profits they might generate to their owners, managers, or others; and third, these entities are not controlled by government. To be sure, NPIs form one subset of units and activities generally considered in scope of the civil society sector, but this subset represents most of the economic weight of the civil society activities.

The recognition of the NPIs within the SNA is a relatively new phenomenon, introduced in 2003 with the publication of the *Handbook on Nonprofit Institutions in the System of National Accounts*, which established the first officially sanctioned procedure for capturing the work of non-profit organizations in national economic statistics, followed with publication of the *ILO Manual on the Measurement of Volunteer Work* in 2011, the first internationally sanctioned tool for gathering official data on the amount, character, and value of volunteering. It represents a major breakthrough in economic statistics that until recently used accounting rules making the NPIs virtually invisible.[1] This breakthrough was possible, for a large part, thanks to the groundbreaking work of the Johns Hopkins Center for Civil Society Studies (CCSS) that pioneered a systematic cross-national "mapping" of the civil society sector organizations, known as the Comparative Nonprofit Sector Project (CNP). As will be described below, the CCSS developed a methodology for identifying NPIs independently of their country-specific attributes (such as a legal status or forms of financing), and quantitative measurement of their key dimensions, such as labor input, including volunteer work, operating expenditures, and revenue sources and streams. After being successfully implemented in over 40 countries, this methodology has been adapted by the United Nations Statistics Division and the International Labor Organization as the official methodology of defining and measuring NPIs and volunteer work, described above.

The purpose of this chapter is to explain how the CNP project went about this task in the hope that this might provide useful insights into the social economy sector in capturing this broader set of entities in international statistics as well. To do so, we first examine the project's basic approach, including the criteria that guided this approach, describe how this facet of the social economy/civil society sector that it focused on was conceptualized, and the basic methodology it employed. We then turn to a number of lessons that this approach might hold for others seeking to extend this analysis into broader segments of the social economy or civil society domains.

1. The John Hopkins Comparative Nonprofit Sector Project: the approach

The Johns Hopkins Comparative Nonprofit Sector Project (CNP), developed by the Johns Hopkins Center for Civil Society Studies (JHU/CCSS) and a team of researchers around the world, was conceived as the first systematic, empirical study that developed and implemented a cross nationally comparable conceptualization and measurement of NPIs. The CNP is still active, and new countries continue to participate. In this section we identify the basic criteria that guide this inquiry, the definition of the range of social phenomena considered to be in-scope for the inquiry, and the basic methodology deployed.

---

[1]   For a detailed discussion on the treatment of NPIs in the SNA see chapter by Édith Archambault.

1.1. Key criteria

The Johns Hopkins CNP first formulated a set of criteria to guide its work. In particular, these criteria sought to ensure an approach that is:

- *Comparative*, covering countries at different levels of development and with a wide assortment of religious, cultural, and political traditions;

- *Systematic*, utilizing a common definition of the entities to be included and a common classification system for differentiating among them;

- *Collaborative*, relying extensively on local analysts to root our definitions and analysis in the solid ground of local knowledge and ensure the local experience to carry the work forward in the future. Accordingly, we recruited a principal Local Associate in each country to assist us in all phases of project work and met with these associates regularly through the life of the project to formulate research strategies, review progress, and fine-tune the approach;

- *Consultative*, involving the active participation of local civil society activists, government leaders, the press, and the business community in order to further ensure that the work in each country was responsive to the particular conditions of the country and that the results could be understood and disseminated locally;

- *Empirical and objective*, moving wherever possible beyond subjective impressions to develop a body of reasonably solid empirical data on the social reality being examined; and

- *Institutionalizable*, using procedures and concepts sufficiently consistent with international statistical usage to permit them to be incorporated into official statistical data-gathering and reporting.

1.2. Conceptualization of the sector

Comparison is only possible if reasonable care is taken in specifying what is to be compared. Given the conceptual ambiguity, lack of knowledge, and ideological overtones that exist in this field, the task of conceptualizing the sector naturally had to be approached with care. The CNP approached this task from the bottom-up, building up our definition and classification in collaboration with the Project's Local Associates from the actual experiences in project countries. The goal throughout was to formulate a definition that was sufficiently broad to encompass the diverse array of entities embraced within this sector in the varied countries covered, sharp enough to differentiate these entities from those that comprise the market and the state, the two other major sectors into which social life has traditionally been divided, and operationalizable enough to permit the development of cross-country comparative data on this sector and its various components by member-nation statistical agencies.

Out of this process emerged a consensus on five structural-operational features that became what we have termed the "structural operational definition" of the nonprofit sector. Under this definition, the nonprofit sector is composed of entities that are:

- *Organizations*, **i.e., they have some structure and regularity to their operations, whether or not they are formally constituted or legally registered.** This definition embraces informal, i.e., non-registered, groups as well as formally registered ones. The defining question is not whether the group is legally or formally recognized but whether it has some organizational permanence and regularity as reflected in regular meetings, a membership, and a set of procedures for making decisions that participants recognize as legitimate, whether written or embedded in spoken tradition.

- This focus on organizations keeps the approach consistent with international economic statistical systems, which focus on "entities," as noted earlier. It also means, however, that the coverage is narrower than that used by some scholars operating under the "civil society" conceptualization," who extend the concept to include not only informal organizations but also individual forms of social action such as mass movements and demonstrations.

- *Private*, **i.e., they are institutionally separate from the state, even though they may receive support from governmental sources**. This criterion differentiates civil society organizations from government agencies without excluding organizations that receive a significant share of their income from government, as many civil society organizations now do. This criterion differentiates the CNP definition from that used in the System of National Accounts, which, as noted earlier, considers the source of organizational revenue in determining which sector of the economy it should be allocated to.

- *Not profit-distributing*, **i.e., they are not primarily commercial in purpose and do not distribute any profits they may generate to their owners, members, or stockholders.** This criterion differentiates nonprofit institutions from for-profit businesses and thus meets both the *clarity* and *operationalizability* criteria for the CNP approach, since the vast majority of countries utilize such a non-distribution constraint in their legal structures for the sector. It also aligns the CNP definition with existing statistical usage, enhancing the prospect that we could engage the international statistical system to carry on the CNP's data-gathering work. At the same time, this feature excludes many cooperatives and mutuals from the project's focus, at least where such organizations are permitted to distribute any surplus they earn to their members. Mutuals or cooperatives operating under laws prohibiting such distribution are included however.

- *Self-governing*, i.e., **they have their own mechanisms for internal governance, are able to cease operations on their own authority, and are fundamentally in control of their own affairs**. This criterion differentiates nonprofit institutions from subsidiaries or agencies of other legal entities, including units of government;

- *Non-compulsory*, i.e., **membership or participation in them is contingent on an individual's choice or consent, rather than being legally required, or otherwise compulsory**. This criterion is useful in differentiating NPIs from kin-based groups (e.g., extended households, tribes, or castes) whose membership is determined by birth rather than individual consent.

The result is a quite broad definition of the nonprofit sector, encompassing *informal* organizations as well as *formal* organizations; *religious* as well as *secular* organizations; primarily *member-serving* organizations, such as professional associations, as well as public-serving ones. What is more, the definition rests on clearly operationalizable criteria rather than subjective judgments (such as whether the entities are serving the public good) or the legal status of organizations, which often varies among countries.

1.3. Data assembly methodology

The second critical element of the CNP methodology was the formulation of a standard set of methods and quantitative data assembly strategies designed to generate a comprehensive multi-dimensional and cross-nationally comparable measures of NPI activity. Instead of employing the often misleading variable of the number of organizations, the CNP methodology focuses on variables that provided a clearer picture of the levels of actual activity of the entities under examination. To this end, it identified six empirical dimensions allowing objective and cross-nationally comparable measurements of different aspects of the nonprofit sector's activities. These include:

1. number of paid employees, converted to full-time equivalent (FTE) jobs;[2]
2. number of volunteers, converted to full-time equivalent jobs;

---

[2] The conversion to FTE allows a more accurate measurement of labor input than a count of the number of persons, because it accounts for differences in the actual number of hours worked; the problem of part time engagement is particularly severe for volunteers.

3. total amount of expenditures;

4. amount of revenue from government payments;

5. amount of revenue from private philanthropy; and

6. amount of revenue from sales, service fees, membership dues and investments.

To collect the needed data, a set of research protocols was formulated defining the data items being sought and suggesting ways to secure the needed data. The main goal of the data assembly strategy is to produce these six data elements for the entire nonprofit sector in each project country using, to the extent possible, existing administrative and statistical data collected by government agencies. Only when the existing data sources are inadequate or altogether unavailable does the research team commission special surveys to collect new data.

As a further aid in depicting the sector, JHU/CCSS and the CNP Local Associates formulated a classification system for differentiating these entities according to their primary activity. Just as the definition sought to align itself with the international standard to the extent possible, this classification structure similarly adhered closely to the existing International Standard Industrial Classification (ISIC), which is the structure used in general international statistical work. However, because the level of detail in ISIC, Rev. 3, the one in force at the time of the start-up of the project, lacked sufficient detail in the fields of activity in which NPIs tend to cluster, a more detailed classification structure was formulated called the International Classification of Non-profit Organizations (ICNPO). The six quantitative measures were then further distributed into fields in which NPI activity tends to concentrate.

Another critical aspect of the CNP methodology was the consultative and collaborative data assembly and verification process, achieved through the ongoing application of the Delphi method[3] (Linstone and Turoff, 1975). The entire research process, from the formulation of the project's conceptual framework – its working definition, treatment of borderline organizations, classification system – to planning data-collection

---

[3] The Delphi method is a technique aiming at reaching a consensus in a panel of experts through the process of structured communication. This method was first used for military forecasting but it has been widely adopted in many fields. Although there are many variants of this method, all use iterative communication process which begins with experts expressing their professional opinions which are then reviewed and commented by other experts. In the subsequent iterations, the experts are encouraged to revise their original judgments taking into account received feedback. The process continues for a pre-defined number of iterations or until a consensus emerges.

strategies, data collection, data verification and analysis, and to reporting the finals results is based on a consensus of all involved parties. In each country we recruit a principal Local Associate, who leads the research at the national level and acts as liaison between the JHU/CCSS research team and local experts and stakeholders. Local Associates meet regularly throughout the life of the project to formulate research strategies, review progress, and fine-tune the approach. These individuals in turn recruit colleagues to assist in the effort. The result is a project team that has engaged at least 150 local researchers around the world in the development and execution of the project's basic tasks.

This collaborative approach was instrumental in developing, testing, and applying the CNP structural operational definition in the local context of every country covered by the project,[4] and though the Project is now 20 years old, each new country team that joins the CNP tests the application of the operation definition in the local setting, to see how well it captures the civil society organizations at the local level. The main purpose of this exercise is to determine that the proposed concept does not act as the proverbial Procrustean bed in the local context, by either chopping off the entities that most observers would consider a core element of the civil society sector, or by stretching the scope so that it includes entities which are not part of the civil society sector by any stretch of imagination. The fact that the CNP definition excludes or includes certain local organizational forms from the scope of the civil society sector does not automatically mean a poor fit the local context. In most situations, the researchers face borderline cases that either formally meet the definitional criteria, but locally are not considered to be in scope of the civil society sector (e.g., labor unions, political parties, or state churches), or that do not fully meet all the definitional criteria, but locally are considered to be in scope (e.g., certain types of cooperative and mutual help organizations).

This goodness of fit testing procedures have, to date, revealed that the CNP definition fits the local realities rather well, but there are few exceptions. On the one hand, there are institutions that meet the international definition criteria, but locally are not considered to the a part of the civil society sector. These include hospitals and universities in Canada, which locally are considered a part of the public sector, and sports organizations in the UK. On the other hand, there are institutions that are locally considered to be a part of the civil society sector, but do not meet the international definition criteria. Examples include state churches in Switzerland and Scandinavian countries, or territorial self-defense associations in

---

[4]   The results of each test are available in the publications database available at http:// ccss.jhu.edu/ as part of the CNP working paper series.

Denmark. Given the diversity of civil society institutions, the existence of these exceptions is not surprising. On the one hand, it confirms the discriminating capacity of the operational criteria set forth in the CNP conceptualization, effectively allowing identification of units that are in scope of the study regardless of their local treatment. On the other hand, the existence of these exceptions called for a modular approach to measuring NPIs internationally, to avoid the definition becoming the proverbial Procrustean bed which enforces uniformity without regard for individuality. The essence of the modular approach is separate identification in the data of those subsets of institutions that meet the international but not the local criteria, as well as those meeting the local but not the international criteria, so they can easily added to or subtracted from the statistical aggregates to suit a particular focus of inquiry. This modular approach to the presentation of the project's result was an important factor facilitating local "ownership" of these results.

Once the goodness of fit test has been completed in a project country, the collaborative effort is used to develop a data assembly strategy. This involves the identification of the existing data sources and the scope of their coverage, and the development of instruments to supplement missing information, if needed. As the data are being assembled, the interim results are systematically communicated with other researchers involved in the project for internal consistency checks and verification. A crucial part of the data verification process involves cross-national comparisons to spot out-of-range values. This method has proven extremely valuable in testing the quality of the obtained results because it allowed to identify results requiring further study to determine whether they reflected a true abnormality or were the result of human error in the data collection and compilation process.

As a further guarantee, the Project involves the active participation of local civil society activists, government leaders, the press, and the business community in order to further ensure that the work in each country is responsive to the particular conditions of the country and that the results can be understood and disseminated locally. To achieve this, we organize Advisory Committees in each project country and at the international level. These committees review all aspects of the project approach, assist in the interpretation of the results, and help publicize the findings and think through their implications. Altogether, more than 600 nonprofit, philanthropic, government, and business leaders have taken part in the project through these Advisory Committees.

These efforts have so far produced the most extensive body of knowledge on the scale, composition, and financing of civil society institutions in 43 countries representing six continents, diverse cultural traditions, and different levels of economic development.

1.4. Development of official data sources

The success of CNP demonstrated the feasibility collecting comprehensive data on nonprofit institutions. However, the CNP approach had one drawback – labor intensity resulting in high cost – that made its continuation very difficult. The great public demand for the CNP data made it clear that a more permanent fix was needed for institutionalizing the generation of data, thus ensuring continued availability of information on the nonprofit sector. The focus thus turned to finding a way to change official statistical accounting methods.

Both the United Nations Statistics Division (UNSD) and the International Labor Organization (ILO) were fortunately receptive to this effort, and adopted a CNP-type approach for generating official data on the nonprofit sector and volunteering respectively, enlisting the JHU/CCSS to work with an international team of experts to produce official guidance documents for government statistical use. As a result of this effort, in 2003 the UNSD published the *Handbook on Nonprofit Institutions in the System of National Accounts*, which established the first officially sanctioned procedure for capturing the work of non-profit organizations in national economic statistics. In 2011, the ILO followed with publication of the ILO *Manual on the Measurement of Volunteer Work*, the first internationally sanctioned tool for gathering official data on the amount, character, and value of volunteering. By implementing these tools, governments utilize the expertise and resources already in place at government statistical agencies to collect the data, thereby minimizing the costs otherwise associated with large-scale national surveys and research efforts, while producing a wealth of new data on the civil society sector and volunteerism which will allow countries and stakeholders to compare results in similar and disparate nations, providing hard data to assess their efforts and to better direct their work.

The UN *NPI Handbook* specifies the same five operational criteria for defining non-profit organizations and produces a full set of financial variables on the sector in a so-called "satellite account on nonprofit institutions,"[5] and specifies as well additional data elements that are of

---

[5] The terms 'nonprofit organization or sector,' 'nonprofit institutions (NPIs),' and 'civil society organizations (CSOs) or sector' are used interchangeably here. Although there are subtle differences among these terms, they are all used extensively in the literature to refer to a common array of entities, and they will be used here to refer to a particular array of entities that fit the definition presented in this chapter. The term NPIs is used more commonly in the context of official national data because it is the term that has been adopted by the United Nations System of National Accounts. Not used is the term 'civil society' as a noun because this often embraces far more than the entities covered by the definition presented in this chapter.

particular relevance for nonprofit institutions. These include the number of employees and volunteers, the monetary value of volunteer input, the value of non-market output of "market" NPIs,[6] as well as the amount of government payments and reimbursements included in proceeds received for the sale of goods and services, as well as the breaking down the received grants and donations, or "transfers" in SNA terminology, by their source (government vs. private).

Whereas the CNP project is implemented by private researchers working together with the JHU/CCSS, the implementation of the *UN NPI Handbook* is carried by government statistical agencies in the course of performing their statutory duties (though many have collaborated with CNP Local Associates and JHU-based team to carry out their work). The main benefit of this approach is that NPI satellite accounts have the status of official statistics. JHU/CCSS has launched a global dissemination, technical assistance, and implementation campaign to promote this method of data assembly on the nonprofit sector.

Since the publication of the *UN NPI Handbook* in 2003, 16 countries have produced NPI satellite accounts for at least one year, and eight of these have produced accounts from multiple years (Salamon, 2013).

2. Impacts of the CNP

What has this effort accomplished? What are its lasting contributions? Getting definitive answers to these questions is difficult. Many forces have been at work in our world over the life of this project that have affected the nonprofit sector and its role in national and international life. Separating out the impact of these other factors from the impact of the work is no mean task. What is more, definitive evidence of the impact of our work is often difficult to track down and demonstrate. It is resident in the minds and work of others and only occasionally manifests itself – in an email exchange from a far-off scholar who has read one of our articles or books, in a reference to our work in a speech by a UNDP official, in a resolution of the European Parliament, and in other likely and unlikely places. Despite these difficulties, we believe that this work has had, and will continue to have, quite significant consequences and that this can be demonstrated with concrete, if far from complete, evidence. Five accomplishments in particular seem worthy of note.

---

[6] This represents the value of goods or services that NPIs provide for free or at below-the-market prices. If an NPI is a assigned to the corporations sector in SNA, its non-market output is typically not reported.

2.1. Conceptualizing the nonprofit sector

Perhaps the most significant accomplishment of the work reported here has been its contribution to conceptualizing the nonprofit sector as a definable and researchable arena of social action. Hard though it might be to realize today, when our work began about 25 years ago, there was no agreed upon definition of something called the "nonprofit" or "civil society" sector. The prevailing paradigm pictured a two-sector social reality made up of the market and the state. The idea that there was a third sector between these two with enough commonality to constitute a distinct economic sector was non-existent in most places. Even in places where such institutions clearly existed, the differences among the different types of organizations – hospitals, social service agencies, environmental groups – seemed so marked that the idea of a common sector identity had not yet surfaced. A variety of partial definitions was in vogue in different parts of the world, and often within individual countries. Thus scholars and practitioners in the UK differentiated between "charities" and associations; those in many developing countries used the term "NGO," but excluded philanthropies, private hospitals, private universities, and religiously affiliated assistance agencies. In some countries, cooperatives were included and in others not. Finally, within the statistical community, economic units that were not-for-profit by law such as American universities or hospitals were treated as part of the for-profit business sector or the government sector in economic data depending on whether the preponderance of their revenue came from fees or government grants. What is more, the different parts of this large sector – health, education, development, social services, arts, culture – operated in splendid isolation from each other, rarely acknowledging, let alone acting upon, the significant commonalities they shared.

The basic definition we forged with our associates, and the accompanying ICNPO classification, both of which occasioned a great deal of discussion, have now been widely accepted as the international standard for defining the nonprofit sector. In the process, they have created a language and set of concepts through which people from a wide assortment of vantage points have come to understand this sector and what it embraces. Even those who quarrel with this definition by arguing for the inclusion or exclusion of particular types of organizations begin with this definition as their touchstone.

2.2. Increasing the sector's visibility and credibility

Our CNP project generated some of the earliest, and still the most comprehensive, data on the civil society sector. This boosted the visibility and credibility of the nonprofit sector enormously. Prior to our research, the nonprofit sector was treated as a step-child in the media, in policy debates, and in academia – rarely mentioned and typically considered, if at all, as a small-change operation comprised of committed do-gooders. Our research helped to change this image enormously, demonstrating that the nonprofit sector is an enormous economic force, contributing as much to the Gross Domestic Product of countries as the entire finance, insurance, and real estate industries and considerably more than construction and utilities. This realization has clearly penetrated the consciousness of the media, of policy-makers, and of the sector itself. Evidence of this use of our data and the growing recognition of the sector it has helped trigger is apparent in numerous sources, including major media outlets, European Union policy documents and legislation, United Nations reports and resolutions, and in speeches by major public and international officials. As an official in the Ministry of Social Development in New Zealand with which we have been working put it:

> The CNP project has had a significant impact on the visibility and voice of the non-profit sector in New Zealand – one that needs to be celebrated. Personally, it has been very gratifying to see the illumination of understanding that has resulted from the project. A large range of people – from academics to economists, politicians to social commentators, are realizing the importance of the sector both as an economic force and as an essential pillar of New Zealand society.

2.3. New insights/new conceptualizations

In the course of generating some of the first empirical data on the civil society sector at the global level, our work has also challenged a variety of conventional beliefs about the nonprofit sector and about the broader social and economic context in which it operates. In addition, it has also allowed us to formulate an alternative theory – the social origins theory of civil society development – about the factors that give rise to nonprofit organizations and determine their scope and contours (Salamon *et al.*, 1996 and 2000, which has been further expanded in our late book, Salamon, 2014). These insights and conceptualizations have stimulated a considerable amount of intellectual ferment not only among students of the third sector but also among students of social policy and social life more generally. Among the insights and conceptualizations triggered by our work are these:

- **A major economic force.** The discovery that the nonprofit sector is a significant economic force, with a workforce that reaches 10% or more of the economically active population in many countries, putting it on a par with major industries, such as communications, transportation or construction.

- **Revenue structure.** The realization that philanthropy plays a much smaller part in the financing of nonprofit activity throughout the world than previously believed. Our data reveal that on average only about 12-13% of nonprofit revenue worldwide derives from philanthropy, with earned income (including program service fees, sales, membership dues and investment income) and government support making up the balance.
- **The role of volunteers.** The fact that volunteers comprise more than 40% of the full-time equivalent workforce of the nonprofit sector globally.
- **The myth of the welfare state.** The discovery that the image of the "welfare state" portrayed in much of previous social science analysis and much of popular public discussion is seriously misleading in portraying government as the principal provider of health and welfare services. In fact, our research demonstrated that enormous nonprofit sectors operate in many of what were previously considered to be classic "welfare states." In fact, what really exists in many of these countries – especially countries with a substantial Catholic population – is a "welfare partnership" in which key social welfare services are financed through government but most of the services are actually delivered through private, nonprofit organizations.
- **Debunking the supposed displacement of paid staff by volunteers.** There is a common misconception, especially popular among organized labor, that engaging volunteers displaces paid employment. The data we generated strongly challenges this perception implying that the relationship between volunteers and paid staff is a zero-sum one. Rather, these data suggest a positive relationship between paid staff and volunteers: the more paid staff the more volunteers that appear to be mobilized. This underlines the importance of careful management of the volunteer function as a way to mobilize volunteer workers.
- **Debunking the myth of adverse government – nonprofit sector relationship.** This myth, especially popular in English speaking countries, holds that government and nonprofit institutions are economic competitors and often political adversaries. Not only does this myth underlie much of the popular discourse on nonprofit institutions, but it finds its expression in academic theories. One such theory, known as "government failure" claims that demand for different quality of collective goods provided by government results in limited public funding for such goods due to the fact that public budget requires majority consensus. As the public funding

supposedly dwindles, private charitable institutions fill the void for unmet demand, which results in the growth of the nonprofit sector. Another theory claims the "crowding out" effect of government social spending; according to this theory, government social spending is supposed to muscle out private charity and nonprofit action. Both theories suggest an inverse relationship exists between government and nonprofit organizations – as the state expands, nonprofits should shrink, and vice versa. However, our data show that none of it is true. Countries with most generous social spending, including most Western European countries, are among those with the largest nonprofit sectors. What is more, countries with culturally diverse populations, such as the Netherlands or Canada were able to significantly expand their public budgets funding "collective goods" in cooperation, rather than in competition, with nonprofit institutions. As a result, expansion, not contraction, of public budgets led to the growth of the nonprofit sector. To account for these realities, we formulated an alternative body of theory – a "new governance" theory of the public sector and a "social origins" theory of the nonprofit sector, as noted above.

2.4. Policy impact

By penetrating the consciousness of the media, researchers, and policymakers, the work of our Center has also begun to influence public policy. The following few examples were only included where our research was explicitly mentioned as a driver of the policy:

- Findings from the Johns Hopkins Comparative Nonprofit Sector Project were used extensively to promote a significant new nonprofit law in Japan in the latter 1990s. This law provided a mechanism for formerly unrecognized nonprofit organizations in Japan to secure legal status.

- Findings from this Project also figured prominently in similar efforts to promote more supportive legal arrangements for nonprofits in Italy and helped provide the justification for special ministerial offices on the third sector in both France and Italy.

- The United Nations has relied extensively on our work as a basis for new policies toward the third sector and volunteering. The report of the Secretary General of the UN on progress on the mandates adopted as part of the 2001 Year of the Volunteer made explicit reference to this work and urged further progress on it.

- The European Parliament passed a major resolution urging more supportive European policies toward the nonprofit sector and volunteering that cited our work extensively as a crucial part of the justification.

- The Government of Norway has adopted a comprehensive policy of support to the third sector also based extensively on our work and that of our associates in Norway. As part of this policy, the Government of Norway has implemented the *UN NPI Handbook* and produced "satellite accounts" on the nonprofit sector on a regular basis.

- India's last five-year plan committed the country to a set of policies supportive of nonprofit organizations in important part based on data we were able to provide. As part of this also, the government carried out a comprehensive census of some 3 million nonprofit organizations and developed a satellite account on the nonprofit sector with assistance from our Center.

2.5. Building a scholarly community

Finally, in addition to its substantive accomplishments, our work also literally helped to create an international scholarly community in the field of nonprofit studies and philanthropy. When this work began, scholarly attention to this set of institutions around the world was very limited. Few researchers were working on the topic and those that were had little visibility or credibility. What is more, they were largely working in isolation.

Because of our decision to rely on Local Associates in the project countries and to utilize a collaborative approach to structure an agreed-upon framework for analysis, the project established strong bonds among scholars that formerly had no connections and a sense of common objectives that took cognizance of regional and national perspectives. Altogether over the life of this project a minimum of 200 researchers were actively engaged in the effort. This included the Principal Local Associate in each of the 40+ countries covered plus the three or four colleagues that each of them also engaged in the work. Beyond this, the Project mobilized another 600-800 individuals who took part in the Advisory Committees the project formed in each country to review results, assess definitions, and generally participate in the analytical process.

The International Society for Third-Sector Research, the academic society representing the third-sector research community internationally, began to a significant extent as an outgrowth of the Johns Hopkins Comparative Nonprofit Sector Project. The Society's first President was the Project's Local Associate in Israel. Its first Vice President was the Project's Director. And other key members of the Board had links to the Project. Many of the leading nonprofit scholars of today got their start in this field through the Hopkins project.

With the launch of the United Nations *Handbook on Nonprofit Institutions in the System of National Accounts*, this process is now repeating itself at the level of national statistical agencies. An entire new network of nonprofit institution experts is now being formed within the staffs of national statistical agencies as these staffs take on the job of implementing this *Handbook.*

Over the long run, the mobilization of this exceptional pool of talent for focused attention to the nonprofit sector and philanthropy may represent the most lasting contribution of all of the work described here, for it is the energies and interest of these scholars and activists that will sustain the nonprofit sector, and the popular and academic attention it needs, into the future.

3. Lessons learned from the CNP

As the previous sections have demonstrated, JHU/CCSS has extensive experience designing and implementing descriptive quantitative studies in multiple countries around the world and has worked with government statistical offices to advise them in identifying NPIs in the statistical records of the SNA. What lessons then can be drawn from the more than 20 years this project has been underway, and how can this experience advance future descriptive research on civil society institutions?

Lesson 1. Work within existing structures. There are more data on civil society organizations hidden in official statistics than is commonly believed, at least in countries with at least moderate levels of statistical capability. Until very recently, nonprofits and civil society institutions were virtually absent from officially-produced statistical data. Some of the reasons for this absence include the lack of a commonly accepted definition, a belief that these institutions do not carry any significant economic weight, and that until recently there was no requirement to separately identify them in the SNA statistical records. This, in turn reinforced the mistaken belief that little if any economic data on these institutions exist.

In reality, however, the opposite is true, especially in the OECD and other developed countries. Most countries compile extensive administrative registers of virtually all economic units operating within their jurisdictions and conduct regular surveys and censuses of their economic activities. These registers and surveys cover nonprofit and civil society organizations as well. The only problem is that they do not identify them as such, which is a problem the *UN NPI Handbook* corrects.

The success of the CNP approach in assembling descriptive data on nonprofits in nearly 50 countries demonstrates the usefulness of working within the existing structures to identify relevant data.

Lesson 2. Identification of nonprofit and civil society institutions in the existing statistical records is feasible. Not only do the data exist, these organizations can be separately identified. As a rule, most existing administrative and statistical records do not yet have a "nonprofit flag," that is, a data element indicating whether a particular institutional unit is or is not a nonprofit institution. But even if such "flags" exist, their usefulness may be limited by the fact that the definition of the nonprofit status used to create this flag may not correspond to that in the *UN NPI Handbook*. The most successful approach to overcoming this problem involves applying the operational criteria specified in the NPI definition to relatively homogenous groups of entities that can be easily identified in the existing data.

The procedure to identify NPIs starts with identification of groups at the lowest possible level of aggregation. Examples of such groups may include institutional units that have a specific legal status, such as a specific type of exemption from taxes, or associations, cooperatives, mutuals, foundations, non-stock corporations, units in certain activity fields combined with other information such as ownership type (private vs. public). The next step involves a test, typically employing some variation of the Delphi method to determine whether the entities in each group meet all of the operational criteria specified in the definition. An alternative approach involves matching the existing administrative or statistical microdata against lists of units that are known to meet the operational criteria.

The application of these approaches led to a successful identification of great majorities of NPIs in most countries covered by the JHU/CCSS projects.

Lesson 3. Measurement of some data elements will require the development of specialized tools. In the case of the CNP approach, it was generally not possibly to ferret out data on the amount and character of volunteer work from the existing systems. Volunteers account, on average, for about 40% of the total NPI labor force, and the UN *NPI Handbook* requires including their number and the value of their contribution in NPI satellite accounts. However, this requirement proved to be extremely challenging to implement due to very limited availability of suitable data. With the exception of a handful of developed countries, information on volunteering is not covered in official statistics. The privately sponsored surveys that attempt to measure the scale of volunteering use relatively small samples, diverse, often-incomparable, methodologies, and widely differing definitions. Consequently, the results they produce vary wildly from one measurement to another (Salamon *et al.*, 2011). This poses a serious question about the reliability of these results and disqualifies them from being included in NPI satellite accounts, which must adhere to rigorous quality standards.

As described above, to address this problem, the JHU/CCSS partnered with ILO to develop internationally accepted standards for measuring volunteer contribution, outlined in *ILO Manual on the Measurement of Volunteer Work* (Salamon *et al.*, 2011, 2013). Once adopted by national statistical agencies, this new *Manual* thus promises to significantly impact the data available on volunteer work throughout the world, and to resolve many of the measurement issues that have long impeded the type of systematic, cross-national measurement of the scale and economic value of volunteering.

Lesson 4. The value of comparative approach. The nature of the CNP project carried an inherent danger of veering widely off course without even knowing it. Since the size and distribution of nonprofit activities were previously unknown in each of the project countries, there was no natural benchmark or reference point to guide the data assembly process. Since statistical data are inherently susceptible to selection bias and various measurement errors, it was imperative to develop a verification procedure to test the quality of the obtained results and spot potential problems.

Cross-national comparisons of the interim results proved extremely valuable in this task. Although no two countries are identical, between-country differences fall within a certain range for various group of countries. For example, per capita income in most developed countries in typical year during our study (1995-2008) falls with the range of $10k-$25k. In this light, data suggesting that per capita income in a particular country fall outside that range begs the question whether this result is real or a result of a statistical error, and prompts additional scrutiny to verify this result.

Following this logic, the interim results assembled in each CNP project country were compared, in various ways, with those from other countries. If wide variations in these results were detected, the JHU team worked with the Local Associates to re-examine the quality of the data and the data assembly methodology and form a consensus about a likely explanation of these inconsistencies. The default "null hypothesis" tested by this approach was that no significant variation among the reference countries exist, and observed deviations are an artifact of methodological procedures or poor data reliability. If this "null hypothesis" was confirmed by the critical examination of methodological procedures or data quality, a new more accurate estimate was developed. If the "null hypothesis" was rejected, i.e., if the methodology and data were found to be sound, we concluded that the variations were real and needed to be explained by social, economic, or historical factors.

This procedure allowed an effective evaluation of the quality of the obtained results and correction of biased or erroneous results, if possible.

Lesson 5. Collaboration among stakeholders is the key to successful data assembly. Assembling comprehensive cross-national data on nonprofit institutions is more of an art than science, as it requires making multiple judgment calls throughout the entire process, from identifying potential data sources, to designing methods of identifying NPIs in these sources, producing the actual estimates, and evaluating the quality of these estimates. This process requires collaboration among three types of actors:

1. individuals with expertise in national accounting, legal and organizational aspects of nonprofit activity, and statistical data systems;
2. government agencies administering different data sources and managing the assembly of national accounts; and
3. stakeholders who will be using the data, including policy makers, the media and research community and leadership of nonprofit institutions.

Although the participation and effective collaboration among these three groups of actors is essential for the successful assembly of data, securing this participation and collaboration can be very challenging. First and foremost, government statistical agencies are not free to decide which data to produce and release, but rather respond to specific public demands for data. Such public demands typically come from the political or administrative leadership of the country and, indirectly, from various constituent groups, such as the business sector, the media, the academia, or civil society. Therefore, establishing public demand for reliable data on the nonprofit sector is an essential first step in the process of assembling NPI satellite accounts.

However, public demand alone is not sufficient. The government agencies that will be responsible for generating the accounts must be convinced that they have the necessary organizational and methodological capacities to undertake this task. This may require securing additional funding, either public or private, and providing additional technical expertise on NPIs and their treatment in statistical data systems. At this point identifying potential funding sources and recruiting experts is necessary. It is also beneficial to hire a coordinator to maintain regular and effective communication among experts, statistical agencies, and other stakeholders and monitor the progress of work.

JHU/CCSS successfully utilized this model of collaboration in all comparative project countries, as well as in most countries that implemented NPI satellite accounts.

Lesson 6. Numbers do not speak for themselves. Successfully assembling the data on the NPI sector is the final goal of statistical agencies and experts collaborating with them, but for the stakeholders that seek to use these figures to achieve a policy goal, it is important to make these data public.

The responsibility of data producers, such as government statistical agencies, ends when the data is released to the public domain. From that point, finding the data becomes the responsibility of their users. In case of the routinely released data, most users are well aware of their existences, times of release and the location where they can be found. However, the public may not be aware of data that are released for the first time or sporadically, such as NPI satellite accounts. The situation may be further complicated by the fact that the great volume of statistical data produced in OECD and other developed countries may simply "drown out" the NPI satellite accounts.

Therefore, it is beneficial to take special efforts to increase public awareness of the newly available data on civil society organizations. To that end JHU/CCSS encourages the organization of dissemination events – co-hosted by the statistical agencies and local stakeholders where possible – to release the results at the national level. Ideally, the final report would present the national data in an easy to understand format and would include a section placing the national results in an international context. The event itself typically involves the participation of various stakeholder groups, such as policy makers, the media, the academic community, representatives of statistical agencies that produced the data, civil society leaders, and representatives of international organizations, such as UNDP or ILO. Where the inclusion of international data is not possible, a follow on companion document is produced.

Lesson 7. Continuous monitoring is needed to ensure that implementation produces comparable data and to keep pace with updates that are made to the international statistical systems. The final step in public dissemination of the NPI data involves making them available to the international community. This task is complicated by a number of factors, such as language barriers that make finding NPI data on official web pages difficult for foreigners, lack of standard formats for presenting these data, and the lack of international publicity. Further, the existence of the *UN NPI Handbook* and the *ILO Volunteer Manual* alone does not ensure their consistent application. As a result, the resulting data can be difficult to find and compare.

To remedy this, JHU/CCSS has assumed the role of an international clearinghouse for the available NPI satellite accounts data, and in this capacity carries out the following tasks:

1. Regular monitoring of the NPI satellite account implementation process world-wide;
2. Obtaining the released NPI satellite account data from the agencies that produced them;
3. Converting these data into a standard format;
4. Carrying out the cross-national comparison of the available data; and
5. Production and international release of the findings of these comparative analyses.

Additionally, regular updates are being made to the international statistical systems that impact the measurement of NPIs and volunteer work – the System of National Accounts was revised in 2008 (UN, 2009) and the ILO's *International Conference of Labor Statisticians* took up the question of including volunteering as an official form of work in its 2013 conference. Accordingly, JHU/CCSS has remained involved to identify how the changes affect the measurement of NPIs and volunteering, to lobby for their inclusion and enhancement in the revised versions of the documents, and to modify its guidance documents as needed. Thus, when the System of National Accounts was revised in 2008, JHU/CCSS successfully lobbied for the separate identification of NPIs in the institutional sectors and the inclusion of a separate chapter on the treatment of NPIs (Chapter 23), and has since agreed to revise the *UN NPI Handbook* as a result. The revised *UN NPI Handbook* will take account of changes made to the SNA, especially the recommended subsectoring of NPIs in the core accounts, will clarify new guidelines for classifying NPIs to the government account, and will clarify guidelines for classifying NPIs taking account of updates that have been made to the International Standard Industrial Classification (ISIC, Rev. 4). Additionally, the *UN NPI Handbook* will acknowledge the broader conceptualizations of the Third Sector that are underway in many places and will need to be considered in future developments of official statistics, including measures of cooperatives and mutual societies, and social ventures or social enterprises. JHU/CCSS was included in ILO discussions around the revision of the definition of work in advance of the 19th *International Conference of Labor Statisticians*, held in October 2013, which ultimately resolved that volunteering should be included as an official form of work for regular measurement by countries (ICLS 2013).

Conclusion

This chapter provides an overview of the criteria that guide the JHU/CCSS approach for the measurement and assembly of data on nonprofit institutions and volunteering. The JHU/CCSS engages in two types of research activity: the assembly of quantitative macro-economic data on NPIs in a number of different countries, and the analytic interpretation of these data by integrating them into the internationally accepted system of national accounting. The main purpose of these activities is to produce an empirical description of the NPI sector in both national and international contexts. This differentiates it from other international projects, such as Civicus Civil Society Index or USAID's NGO Sustainability Index, whose main goal is qualitative evaluation against normative standards.

The CNP introduced two major innovations that proved very successful for assembling the most comprehensive cross-national NPI data set to date. The first innovation was the introduction of a standard, internationally "transportable" operational definition of the nonprofit sector that allows for the empirical identification of NPIs in various countries (Salamon and Anheier, 1992). The second innovation was the development of standard data assembly methodology that relies mainly on the existing administrative records and statistics. The essential part of this methodology is systematic collaboration among three types of actors: experts, government statistical agencies, and civil society stakeholders.

The approach developed by the CNP has been subsequently institutionalized in the assembly of NPI satellite accounts within the SNA framework. It constitutes a major building block within a broader statistical system describing macro-economic dimensions of the civil society sector, which is complementary to the satellite accounts on social economy that cover mainly cooperatives and mutuals operating in the market.

JHU/CCSS gained significant experience from implementing this methodology in nearly 50 countries which can provide important insights into current efforts to conceptualize and measure the size and scope of the social economy. Among these are that working within existing structures can lead to more available data than what is commonly believed to exist, though some data elements will require the development of specialized tools. However, their measurement will not alone ensure the successful development of these data. Collaboration among government and civil society actors and continuous monitoring are the key to successful data assembly and dissemination.

References

ICLS, *Resolution I: Resolution concerning statistics of work, employment and labor underutilization*, International Conference of Labor Statisticians, 2013, http://www.ilo.org/wcmsp5/groups/public/--dgreports/--stat/documents/normativeinstrument/wcms_230304.pdf.

Linstone, H. A. and Turoff, M., *The Delphi Method: Techniques and Applications*, Reading, Mass., Addison-Wesley, 1975.

Salamon, L. M., "The Rise of the Nonprofit Sector," *Foreign Affairs*, Vol. 74, No. 3 (July/August), 1994, https://www.foreignaffairs.com/articles/1994-07-01/rise-nonprofit-sector.

Salamon, L. M. (ed.), *New Frontiers of Philanthropy: A Guide to the New Tools and New Actors that are Reshaping Global Philanthropy and Social Investing*, New York, Oxford University Press, 2014.

Salamon, L. M. and Anheier, H. K., "In Search of the Nonprofit Sector: I. The Question of Definitions," *Voluntas. International Journal of Voluntary and Nonprofit Organizations*, Vol. 3, No. 2, November 1992, pp. 125-151.

——, "Social Origins of Civil Society: Explaining the Nonprofit Sector Cross-Nationally," *Comparative Nonprofit Sector Working Paper*, No. 22, Baltimore, The Johns Hopkins Center for Civil Society Studies, 1996.

Salamon, L. M., Sokolowski, S. W. and Anheier, H. K., "Social Origins of Civil Society: An Overview," *Comparative Nonprofit Sector Working Paper*, No. 38, Baltimore, The Johns Hopkins Center for Civil Society Studies, 2000.

Salamon, L. M., Sokolowski, S. W. and Haddock, M. A., "Measuring the Economic Value of Volunteer Work Globally: Concepts, Estimates, and a Roadmap to the Future," *Annals of Public and Cooperative Economics*, 2011, Vol. 82, No. 3, pp. 217-252.

Salamon, L. M., Sokolowski, S. W., Haddock, M. A. and Tice, H. S., *The State of Global Civil Society and Volunteering*, Baltimore, Johns Hopkins University, Center for Civil Society Studies, 2013.

United Nations, *Handbook on Nonprofit Institutions in the System of National Accounts*, New York, United Nations, 2003.

United Nations, *System of National Accounts 2008*, New York, United Nations, 2009.

So, What Does a Social Economy Enterprise "Produce"?

Sybille MERTENS

*Associate professor, HEC-Management School,
Université de Liège, Belgium*

Michel MARÉE

*Senior researcher, Center for Social Economy,
HEC-Management School, Université de Liège, Belgium*

Introduction

Whatever definition is retained, it is generally accepted that a social economy enterprise is an organization that can be considered an enterprise. By enterprise, we mean any entity designed to produce or distribute goods or services, regardless of its legal status or method of financing.[1]

A social economy enterprise stands out in comparison with other private providers of goods and services because it is managed according to non-capitalist objectives. This is illustrated by three characteristic traits: the social aim of the activity, the constraint on the profit distribution and democratic methods of governance (Defourny and Nyssens 2011, Mertens and Marée 2010, Barea and Monzón 2006, Bouchard *et al.*, 2008). These characteristics have two important effects for the ideas we will be presenting. First, in much of the world, social economy enterprises often have specific legal forms, generally of an associative or cooperative nature, that offer these organizations an appropriate institutional framework in which to carry out their production activities. Second, the pursuit of a social aim very often implies relying at least partially upon non-market

---

[1] We use the notion of enterprise as it is defined within the framework of European law (Hoffner ruling at the Court of Justice of the European Community of April 23, 1991 – aff. C-41/9O).

resources to cover production costs (public funding, private donations, volunteering).

These two differences have important consequences in terms of quantification: whether to generate management indicators (profitability ratios, structure ratios, etc.) or to create statistics on a macroeconomic level, conventional measurements often prove to be poorly adapted for providing an accurate quantitative understanding of what a social economy enterprise produces.

There are two reasons for this. First, conventional measurements were designed for enterprises with a capitalist aim and seem obsolete in the context of a social economy enterprise. Let us take "return on investment" (ROI) for example, intended to measure the rate of return of the funds invested in an enterprise by evaluating the relevance of these investments in a cost/benefit perspective. Does this measure apply uncompromisingly to social economy enterprises? It is easy to imagine that the answer is negative; indeed, it seems ill-advised to analyze the return of a social economy enterprise by limiting the analysis to its financial performance, as has incidentally been illustrated by recent attempts to surpass the limitations of conventional measurement for defining "social" return on investment (SROI).[2]

Second, and this is linked to the previous point, the production activity of an enterprise or its value added, that is to say its contribution to the gross national product (GNP), are generally measured using information delivered by the market. However, the production of a social economy enterprise cannot always be totally evaluated by the market. Indeed, the pursuit of a social purpose sometimes leads to the providing of non-market goods or services, without requiring a payment in return that completely covers the production costs, notably because one does not want to reserve access and consumption solely to those who have the means and the motivation to pay this price. Furthermore, consumption (and production) of certain goods and services often generates positive effects on other people than the direct consumers. These effects do not however give rise to a payment through market mechanism. On the other hand, they justify the mobilization of other so-called non-market resources such as public funding, private donations, membership fees, or volunteer work. These resources are mobilizable by social economy enterprises because they produce common goods (recognized as such and financed at least in part by public institutions) or positive externalities recognized by private agents (Santos, 2012; Young, 2007). The market alone seems therefore to be an insufficient mechanism for revealing the complexity of production

---

[2] See notably www.thesroinetwork.org.

of a social economy enterprise. Can a social economy enterprise then be evaluated through a result expressed in terms of monetary value or volume, as in the case of the capitalist enterprise of which the bulk of production is exchanged on the market?

The present chapter attempts to respond to this second question. We indeed wish to demonstrate that because of its complexity, the accurate evaluation of the production of social economy enterprise is hindered by a major conceptual and theoretical problem, particularly when it comes to seeking a monetary measurement of it. Our demonstration is articulated into three steps. We first review (section 1) how the production of the social economy enterprise is now taken into account by the conventions of the national accounting. We later illustrate (in section 2) that it is necessary to introduce a notion of "broadened production" if we want to take into account all the dimensions of what the social economy enterprise really produces. Finally, we conclude (section 3) by showing how this "broadened production" can unfortunately not be the object of a unique monetary measurement, and that it is therefore necessary to let go of the idea that it would be possible to measure the actual contribution of the social economy enterprise to the gross domestic product. We instead plea for recognition of the complexity of the production activity of social economy enterprise and formulate propositions that support the measurement of this production within another framework.

1. Social economy enterprise in today's national accounting

In national accounting, the national production of a country (GNP) is a statistical measurement intended to measure the whole of production resulting from the activity of all economic organizations, whether private enterprises, households, public administrations or nonprofit organizations. The GNP is in fact obtained from three equivalent approaches (European Commission *et al.*, 2009): the sum of the value added generated, the sum of the primary revenues distributed, and the sum of the final expenditures of the economic agents. The first approach, to which we limit ourselves in this work, consists therefore of adding up, for all production units, the total value actually created, that is to say the difference between the value of what was produced (output) and the value of what was consumed in the production process (input or intermediate purchases).

In this calculation, the notion of value is founded, in practice, on the price. In market economy, the production of an enterprise indeed gives rise to an exchange given value by a price, the production value being therefore nothing more than the amount obtained by multiplying the quantities exchanged by the unitary price. By deducting from that amount the intermediate purchases given value by market prices, we obtain the value added of the enterprise, that is to say its contribution to the GNP. This reasoning can easily apply to the enterprise whose production is sold out on the market. But what about the enterprises whose activity consists in part or in whole of providing goods and services outside of the market?

This question has already been asked when national accountants have attempted to assess the value of public services: what is the value of an education service rendered by a school when this service is offered for free? The answer national accounting offers has become a classical one: in the absence of a price, the value of non-market production is measured by what it costs (European Commission *et al.*, 2009). Having deducted intermediate purchases, this value is therefore essentially reduced to the compensation for employees.

And what about an enterprise whose activity can be qualified as hybrid, that is to say made up both of the selling of products on the market and the providing of a non-market service? Such a configuration is not rare and is precisely observed in many social economy enterprises. As an example we will use the case of a work-integration social enterprise that hosts and follows up on vulnerable individuals (non-market production), while providing them work within the framework of a recuperation and recycling activity (market production). For such a case, national auditing relies upon a simple criterion (United Nations, 2003): if the product of sales reaches and surpasses 50% of the total common resources (also include donations, membership fees, public subsidies, etc.), the conventional calculation of value added applies (value of the market production less intermediate purchases); otherwise, the activity is considered as non-market and, as is the case for public services, the cost method is used.

Satellite accounts[3] constitute a good illustration of the aforementioned conventions. The satellite account of nonprofit institutions indeed includes organizations whose value added is calculated from their sales, but also organizations that, like public services, are attributed a value added based on their production costs.[4] The following table indicates, for three countries that establish a satellite account for nonprofit institutions, the contribution calculated in this way by the organizations for the GNP.[5]

---

[3] See the chapter by Fecher and Ben Sedrine-Lejeune.

[4] It should be noted however that the *Handbook on Nonprofit Institutions in the System of National Accounts* recommends using a third option in certain tables of the satellite account of nonprofit organizations: the production should be evaluated based on the highest amount: either the costs or the market resources (United Nations, 2003, 4.80-4.87). In practice, this version of the tables is almost never published.

[5] In France, all nonprofit organizations mainly funded by public subsidies are attributed a value added based on costs.

Table 1. *Value added of the nonprofit institutions (at regular prices)*

Country	Amount		% of the GNP
Belgium (2010)	19,712.2	million €	5.5%
Canada (2008)	35,600	million $ CAN	2.4%
	100,700	million $ CAN (*)	7.1%
France (2002)	45,771	million €	2.9%

(*) Including hospitals, universities and colleges.
Sources: Belgium: ICN 2012; Canada: Statistics Canada 2009; France: Kaminski 2006.

A first limitation of the aforementioned approach is its heterogeneous character because the same method of calculation of value added is not applied to all organizations, except in France. A second limitation results from the underestimation of the value added of the organizations that are considered as market producers (Mertens, 2002). Indeed, for these organizations, production is calculated solely based on sales, while most of them also have non-market production financed with the help of other resources (private donations, subsidies, etc.). Let us take the previous example of the work-integration enterprise and assume that 55% of the resources come from the market (the sale of recycled products), the rest being made up of donations and public transfers. In this example, for national accounting, production is limited to sales of recycled products; it therefore does not take into account training and follow-up activities, etc. carried out within the enterprise thanks to non-market resources.

Aside from these limitations, it is more fundamental to question the pertinence of the method of calculation of non-market production in national accounting. Indeed, whether in relation to a public service or a social economy enterprise, we wish to demonstrate that this production includes dimensions that cannot be evaluated through accounting data, consequently resulting in a marked underestimation of the corresponding value added.

2. For broadening the notion of production of social economy enterprises

Although the reasoning we propose applies to any public service and to any social economy enterprise, it is convenient to continue to use the emblematic example of the work-integration enterprise. This type of enterprise contributes to socio-professional reintegration of workers marginalized on the job market through a commercial activity in the production of goods or services (recuperation and recycling of residual materials, building renovation, organic vegetable growing, etc.). Sales made by this enterprise express the value of its market production. However, as we have seen, this production is only one aspect of the activity. Part of this activity indeed consists of offering a training service and social support to its employees. This service eludes the market exchange and its financing is ensured by resources of another nature (public funding, private donations). The value of the turnover, obtained through the market price, obviously does not express this non-market component of the activity.

However, more fundamentally still, this reintegration mission has significant effects or "impacts,"[6] first and foremost on employees (higher qualifications, re-socialization), but also on the society in terms of social cohesion, the better functioning of the job market, etc. Once again, the conventional measurement of production, through sales, in no way reflects these individual "benefits" (for reintegrated individuals) and these societal "benefits".

In other words, the activity of a work-integration enterprise includes a significantly larger non-market component, which, in our opinion, implies broadening the notion of production itself in order to take into account certain effects on the beneficiaries and on the society in part or in whole. Even if there are no attempts to redefine the notion of production in an enterprise within the norms of national accounting (European Commission *et al.*, 2009), this initiative seems essential to us if we claim to grasp the actual contribution of the social economy enterprise to the GNP, as long as this contribution is measurable of course. We have already described in detail the way in which such an expansion could be conceptualized (Mertens and Marée, 2013). We will limit ourselves in this work to proposing a synthesis based on figure 1.

Figure 1 – *The "broadened" production of a social economy enterprise*

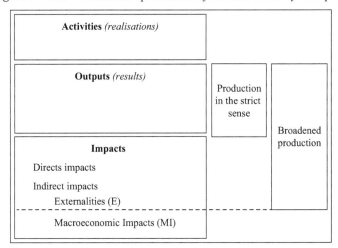

Let us look at the different concepts proposed in figure 1:

The different *realizations* of a social economy enterprise describe its *activities* and correspond to the services rendered within this framework. In the case of an enterprise active in recycling, this is therefore not only the volume of materials collected and the number of products recycled and sold, but also the hours of training and support to workers in a reintegration process.

The realizations give rise to *results* that represent the immediate advantages obtained from the "direct addressees" (also referred to as "beneficiaries" or "users"), that is to say those who directly benefit from the services rendered. In the case of the work-integration enterprise, the direct addressees are obviously the households concerned by the collection of residual materials and the buyers of the recycled products, but also workers benefiting from re-socialization and from better qualifications thanks to the social support offered to them. The results in fact constitute output or *production* in the strict sense of the term. Finally, this *output* generates in turn *direct impacts* on the beneficiaries (improved employability, that is to say a greater aptitude for finding employment) and *indirect impacts* on other economic agents and on the rest of the community.

It is necessary herein to make a fundamental distinction between externalities (E) and macroeconomic impacts (MI) among the indirect impacts. Just like direct impacts, the externalities in the sense that we define them herein have an impact on the utility of the individuals, and in that respect – as shown in figure 1 – they should be considered as constituting a dimension of production of the social economy enterprise. The externalities often have a collective nature, in that they are impacts that concern the group of individuals within a given perimeter. Therefore, insofar as its services contribute to social cohesion, to better use of the resources, or yet to environmental protection, the activity of the work-integration enterprise active in recycling does not only concern the direct beneficiaries (households, clients, and the employees of the enterprise), but rather the whole society.

As for the macroeconomic impacts, they are not to be analyzed in terms of variations in utility of the individuals but in terms of incidence of the activity of the enterprise on other organizations and on macroeconomic figures such as the GNP, the volume of employment, the general level of prices, the public budget, etc. Of course, the macroeconomic impacts of the social economy enterprise play an important role in the evaluation of its role on a socio-economic level. In particular, the economizing of means (cost reduction, avoided expenditures, etc.) represents an important issue. For example, the perspective of reducing the global cost of unemployment public programs constitutes a major argument that heads of work-integration enterprises can present to the public authorities. Not taking into account these types of effects when we try to measure the impacts of work-integration social enterprises would cause a risk for the debate to be cut short. However, these effects do not constitute variations in utility resulting from the activity of the social economy enterprise. So, contrary to the externalities, they cannot be considered as constituent elements of the production (Gadrey, 2002).

In short, based on the chain of effects described in figure 1, we refer to a new concept of broadened production as including output, direct impacts and indirect impacts in the form of externalities.

3. Can production of the social economy enterprise – in the larger sense – be measured and taken into account in the GNP?

After proposing a broadened vision of the production of the social economy enterprise, the question of the monetary measurement of this production inevitably comes up: is it possible to measure the value added corresponding to this broadened production and consequently, to estimate the actual contribution of the social economy enterprise to the GNP? By nature, the direct and indirect impacts that we propose to include in the notion of production do not rely on market mechanisms and, consequently, have a non-market dimension.[7] Their measurement therefore touches on the more general question of the evaluation of the goods that do not transit the market. In these types of circumstances, two major measurement categories exist (Mertens and Marée, 2012): the accounting measurement, which directly lies within the framework of national accounting, and the so-called economic measurement, based on the reconstitution of a shadow price. We will discuss these successively later.

The accounting measurement of broadened production

As its name indicates, accounting measurement of production uses the monetary flows observed to value the production of the enterprise. It is therefore necessary to question whether these flows are correct indicators of the notion of broadened production defined above. By resuming the aforementioned notions, figure 2 shows the problem of accounting measurement of production.

---

[7] This reflection on impacts could also apply to other types of enterprises whose activity sometimes generates (positive or negative) effects beyond those valued by market prices. It seems to us however particularly crucial in the framework of the production of social economy enterprises because the research of these effects is intrinsic to their aim.

Figure 2 – *The accounting measurement of the production of a social economy enterprise*

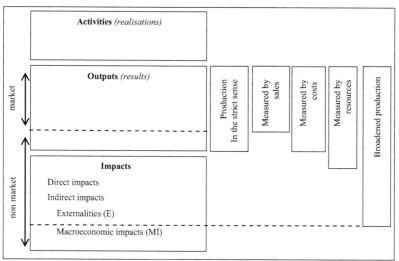

If we settle for using the value of sales, we obtain – as we highlighted above – a measurement of market production. The measurement of production through costs naturally takes into account the non-market portion of the output, or production in the strict sense. But in either case, certain direct impacts (for the beneficiaries) and indirect impacts (for other individuals or for the community) are not taken into account and the production measured is therefore underestimated. In other words, the current norms of national accounting do not allow for an accurate measurement of broadened production.

However, can we not amend the conventions of national accounting by, for example, taking into account the resources side of non-market production, which is neglected in both approaches? Concretely, evaluation "through financing" is directly inspired by the calculation of value added per balance (Mertens, 2002): it indeed consists of making an exhaustive summary of the different resources perceived by the enterprise so as to allow it to ensure its production (sales, membership fees, public funding, donations) and to deduct the cost of intermediate purchases. In this type of evaluation, resources can be interpreted to a certain extent as a request expressed by the financers, the expression of this request being apparent through the mobilization of resources "so the activity can take place".

Although ignored by the national accounting,[8] the evaluation of non-market production through financing or through resources is coherent with that of market production and furthermore corresponds more to a "by the preferences" approach that, as we will point out again later, is advocated by economists to measure the value of a good. This is all the more true if we take into account, in this method of evaluation, all the different types of financing perceived, including "free" contributions, such as donations and volunteer work in particular. As an example we can mention the attempt to calculate a "social value added" by Perrot (2006) that extensively counts monetary resources (sales, public markets, subsidies, employment incentives, memberships, donations and legacy) and non-monetary resources (contributions in kind, free services, service exchanges, exonerations and exemptions, volunteer work).[9] Of course, such an approach requires attributing monetary value to this second category of resources, for example by evaluating volunteer work by analogy with the cost of paid work hours.[10]

If the resources-based approach for valuing a non-market good is therefore more appropriate for evaluating the production of a social economy enterprise, does it not also measure broadened production, that is to say not only the output but also the impacts generated by the production? The answer is partially – and only partially – positive. Indeed, financers of a non-market good are not (necessarily) direct beneficiaries. They are stakeholders who find utility in the provision of the good or service. In other words, by their contribution, they give value to part of the impacts generated, as figure 2 illustrates above.

Unfortunately, the production of the social economy enterprise is only partially measured by the accounting approach because financing of a non-market good generally lies below the optimal level of contribution. Several reasons can explain this apparent incoherence. First of all, the voluntary contributions of individuals (donations, volunteer work, etc.) are generally inferior to the utility that individuals attribute to this production, because of the existence of the well-known phenomenon of the

---

[8] It must however be noted that national accountants consider that certain public subsidies are sales when they take the form of partial or total coverage of the price. In this case, the conventional calculation of value added per balance (sales to the private sector plus subsidies assimilated to sales less intermediate purchases) is similar to the evaluation advocated herein of value added through financing.

[9] Also note the proposition formulated by Deshayes (1988) to surpass the notion of value added in agricultural cooperatives.

[10] See the works carried out under the auspices of the International Labor Organization (ILO) at Johns Hopkins University: http://www.ilo.org/global/publications/books/WCMS_167779/lang--fr/index.htm.

"free rider":[11] Indeed, certain potential contributors renounce financing the good because they think it will be produced anyway. Furthermore, actual contributions are often of a charitable nature that would not allow them to be assimilated to a "price" agreed upon to produce the given good. Indeed, according to Kahneman and Knetsch (1992), as well as many authors that followed, it is all right in many cases to interpret these donations as part of the price to pay to attain "moral satisfaction," rather than as the price of the good itself. They refer to a "warm glow of giving effect" felt by the donor. Even if this affirmation can seem excessive, it is likely that in general, the motivation of moral satisfaction limits voluntary contributions to a level that would be inferior to the amount that preceeded if preferences were fully expressed.

Finally, two arguments can be advanced to explain that public funding does not in any way represent the value of non-market production and does not take into account the whole of the impacts of this production. In the first place, it is common for a hiatus to occur between the wishes of the society and their materialization in terms of political decision and public budget. Indeed, even if individuals recognize the existence of a need, the form in which its financing is perceived (taxes and subsidization) apparently constitute an obstacle to the full expression of their disposition to pay. This obstacle takes on even more magnitude nowadays with the current questioning of the role of public authorities in providing services. In other words, goods for which a value is recognized by the society tend to be under-financed.

Furthermore, the budgetary translation of public support of a non-market activity can be expressed by financing that is not proportional to its recognized value. This discrepancy is due to the connected character of the supply and demand of the non-market good. In the case of the market good, the intensity of the demand is generally expressed by a higher price and therefore also by a higher value. For a non-market good on the other hand, allocated financing is closely linked to the estimated costs for ensuring production. Therefore, an activity to which we attribute significant value because of the positive impacts that it generates may only need relatively little resources for its materialization. The amount of financing is therefore an indication of the value of the good, but does not measure the value itself.

From what precedes, we can only conclude that the accounting approach to the value of non-market goods does not attribute value to all the impacts generated by their production, or it values them in an imperfect

---

[11]   This phenomenon, theorized by Olson (1965), refers to an individual deliberately benefitting from a public good without (or only partially) bearing its cost.

way. In other words, the current tools of national accounting do not make it possible to monetarily measure the broadened production of the social economy enterprise, and consequently its actual contribution to the GNP.

The economic measurement of broadened production

By "economic measurement" of broadened production, we mean a monetary evaluation of the non-market portion of this production founded on the attribution of a shadow price. Because, by definition, there is no market price for a non-market good, we can attempt to replace this absence of price by a method that takes its inspiration from the functioning of the market: starting from the principle that the value of a good depends on individual preferences, it is necessary to attribute a price to the non-market good that corresponds to what individuals would pay if it was proposed on the market. This approach, derived from the problematic of the evaluation of public policy and of major infrastructure works initiated in the 1960s, gave rise to several techniques to evaluate this price (for a detailed presentation, see Mertens and Marée, 2012).

What we can therefore call the "indirect monetary value" of a non-market good is an estimation that is not in fact a direct result of individuals' preferences but that is based on alternative methods inspired by evaluations carried out in the environmental field: cost of replacement, cost of opportunity, avoided expenditures, time-saving, value of human life, etc. A more direct approach would be to resort to the "willingness to pay" (WTP) of the individual: as its name indicates, the idea underlying this measurement is to estimate the value of a non-market good by observing what individuals would be prepared to sacrifice in monetary terms to benefit from this good.[12] In this respect we can identify the WTP based on so-called "revealed" preferences, when it is estimated through the price of substitute or complementary goods, and the WTP based on the "expressed" preferences, when it is obtained directly within the framework of surveys simulating a fictitious market (known as the "contingent valuation method," or CVM).

The aforementioned SROI constitutes a recent application of these techniques to social economy enterprise. It is indeed necessary to express the "social return" of the enterprise in the form of a unique amount by attributing a monetary value to each type of impact generated, while resorting to the different methods for estimating non-market goods whenever possible (ESSEC 2011).

---

[12] Or, alternatively, what individuals would want to perceive in cash ("willingness to accept," or WTA) to *give up* the given good. In the text that follows, we do not make the distinction between these two measurements.

Although attractive in theory, the economic measurement of non-market production is nevertheless met with strong reservations, whatever the framework it is applied to (Mertens and Marée, 2013). First of all, it must be recognized that in practice, resorting to the contingent valuation method (CVM) to measure the WTP of individuals in a SROI analysis has not, to our knowledge, led to any real application. The reason resides for a large part in the fact that, under the guise of simplicity (questioning people on their consent to pay for a service), this method requires carrying out surveys within a representative sample of the population, which is extremely costly in terms of time and energy. It is indeed unrealistic to systematically resort to population samples and to field surveys to evaluate the impacts of a social economy enterprise within different groups. This method of evaluation should in fact be reserved to initiatives of a certain magnitude, such as the evaluation of a public project or the estimation of environmental damages. As for the estimations of the WTP through the price of substitute or complementary goods (revealed preferences), they certainly are more feasible in theory, but they often turn out to be difficult to carry out in practice. Indeed, goods are never perfectly complementary or substitutable, and the estimations obtained this way are consequently only indications of, and not representative of, the value of the good to be estimated.[13]

More fundamentally, we must question the pertinence of the monetary evaluation of non-market goods obtained from preferences (WTP), whether they are revealed or expressed. Is the pretention of economic analysis to be able to measure monetarily actually founded, and is it possible to obtain monetary evaluations that have real meaning based on individual preferences and consent to pay? Among many economists, including those who work within the scope of the third sector and social economy enterprise (see for example Handy, 2003), we observe a favorable attitude toward the WTP approach for the evaluation of non-market goods. But can we keep quiet the fact that the WTP approach has been the objet of much criticism since it was introduced in the environmental sector (see notably Hausman, 1993)? This criticism is not only directed at methodology (for example, concerning the feasibility of collecting representative data in a CVM), but also and especially, as shown below in table 2, at conceptuality, because the criticism is linked to the underlying hypotheses of the method (see Mertens and Marée, 2012, for more details).

---

[13]   The substitute goods for example can rarely be considered as goods that procure services really equivalent to the good to be estimated. The suggestions proposed by the SROI guides are in this respect lend themselves to criticism. Therefore (see ESSEC 2011, p. 36), can we consider that the impact that constitute the improvement of the mental health of the disabled benefiting from the follow-up services of an organization can be evaluated through the cost of the psychological consultations that could be avoided?

The indetermination of individuals' preferences with regard to non-market goods, the indetermination of the variations in utility, the difficulties of taking so-called "non-usage" values into account and, finally, the limitations linked to the simple aggregation of individual preferences, are all arguments in favor of considering that monetary evaluations obtained for non-market goods within the framework of the approach of the WTP are largely subject to caution. Unfortunately, as the presuppositions of the model are often implicit and rarely explicitly presented, its limitations are largely hidden by the appeal of the evaluations that it is supposed to generate. It is therefore not really surprising that the literature on SROI analysis unreservedly advocated resorting to monetary evaluations based on the WTP, as if these raised no debate. In fact, the criticism directed in the last twenty years to monetary evaluations carried out in the environmental sector[14] entirely applies to attempts to evaluate impacts with the help of the WTP as advocated within the framework of SROI.

Table 2. *The conceptual limitations of the WTP approach for non-market goods*

	Hypotheses	Objections	Consequences
1	All individuals maximize their utility function with regard to all states of the world (completeness and transitivity of the preferences)	Bounded rationality of the consumer	Indetermination of preferences
2	The individuals attribute a monetary value to their variations in utility (cardinal utility)	Bounded rationality of the consumer	Indetermination of variations in utility
3	Utility is assimilated to individual well-being	Existence of "non-use values" that cannot be reduced to variations in well-being	Not (or inadequately) taking into account non-use values
4	The value of a good is simply the sum of the individual subjective values	"Bounded awareness" of the consumer: existence of "socially constructed" values distinct from the sum of the individual subjective values	Not (or inadequately) taking into account social values

---

[14] For a recent critical analysis more specifically concerning CVM, see Hausman (2012).

In short, attributing a shadow price to a non-market production presents a number of methodological and conceptual problems that, in our opinion, make this approach inappropriate for measuring the broadened production of a social economy enterprise and therefore its "true" value added.

Conclusions

The analysis that we bring to the definition and the measurement of the production of social economy enterprise leads to an apparent paradox. On one hand, we illustrate that the production of this type of enterprise includes a non-market dimension largely underestimated by national accounting and that it would be necessary to adopt a broadened definition of this production that includes the impacts of the enterprise on its direct beneficiaries but also on the society. On the other hand, having reviewed both ways of measuring non-market production – accounting measurement and the economic approach – we arrive at the conclusion that neither of them can lead to an exhaustive and reliable monetary measurement of the impacts generated by social economy enterprise. In other words, there does not seem to exist an appropriate method of calculation for adequately measuring the contribution of social economy enterprises to the GNP.

This paradox in fact resides within the limitations of the framework of national accounting for helping to understand production in all its dimensions. This framework does indeed not allow for measuring the direct and indirect impacts – whether positive or negative – of a production activity, whenever these impacts do not give rise to accounting flows related to the given activity (with regard to this, we could refer to a "non-internalization" of the impacts). Environmental damages from industry are a good example of negative impacts that are not taken into account in the estimation of an activity's value added and its contribution to the GNP.[15] As far as organizations that generate many positive impacts by their very nature – social economy enterprises, but also, as we should keep in mind, public services – are attributed a value added from national accounting that is not representative of their real societal contribution.

Since precisely defining the production of social economy enterprise is a legitimate concern, it is important to be conscious of the limitations of the measurement tools proposed by national accounting and economic

---

[15] If the GNP does not directly count the externalities (whether positive or negative), these externalities can at least have an *indirect* influence on the GNP in a way that is generally contrary to what was hoped for. For example, water or soil pollution (negative externalities) can give rise to offsetting expenditures that actually increase the GNP.

analysis. This does not mean renouncing the measurement of the value added of social economy enterprises and their contribution to the GNP, whether according to current norms of national accounting or to possible new advances. However, one must avoid interpreting these estimations as measuring the actual production of social economy enterprises.

The notion of "broadened production" we propose indeed implies the definition of another framework than that of national accounting. Concretely, going beyond the conventional notion of value added – monetary amount that measures the contribution to the GNP – it would be necessary to place the measurement of the production of social economy enterprises (and even that of public services) within a complementary procedure that would give up the use of a unique measurement, especially monetary, and would even avoid any notion of "value". Of course, if this suggestion has the merit of being intellectually satisfactory, that would mean an actual near-renunciation to work on this issue on a macroeconomic level.

This is at the core of the trade-off. Either we attempt to be complete and the complexity of the questions at hand – and particularly the definition and the measurement of the impacts generated on individuals – favors a measurement of production founded on an orderly group of indicators that are complementary, quantitative, and qualitative, monetary and non-monetary, and this makes any aggregation attempt impossible.[16] Or we try to work on the macroeconomic level, in which case we must renounce producing data that satisfactorily reflect production of social economy enterprises in the wider understanding. Yet a double path forms between these two opposing paths. First, satellite accounts encourage the development and the use of quantitative non-monetary indicators specific to the industries or to the types of organizations. These indicators can also be used when trying to carry out multi-criteria evaluations of public policy, to compare the production of social economy enterprises over time and to develop international comparisons on this sector.[17] Second, since it seems impossible to acquire reliable indicators that allow us to agglomerate data pertaining to the broadened production of social economy enterprises

---

[16] Many tools that work along the principles promoted herein, and that can be linked to the domain of cost-effectiveness analysis, were developed these last few years to attempt to measure the non-financial "performance" of enterprises with the help of different indicators: "Triple Bottom Line" (TBL), "Impact Reporting and Investment Standards" (IRIS), "Social Accounting Network" (SAN), "Framework and Global Reporting Initiative" (GRI), etc. Mouchamps (2014) offers an analysis of the applicability of these tools to social enterprise.

[17] With regard to this, we refer the reader to the chapters written by Archambault and by Artis, Bouchard and Rousselière.

and therefore to develop a subtle macroeconomic analysis of this sector, it is instead necessary to introduce a reflexive or deliberative approach (between the stakeholders) in order to reach political consensus on the measurement of broadened production of social economy enterprises. A multi-criterion approach can certainly contribute to clarifying these reflections and debates.

References

Barea, J. and Monzón, J. L., *Manuel pour l'établissement des comptes satellites des entreprises de l'économie sociale: coopératives et mutuelles*, Brussels, CIRIEC, 2006.

Bouchard, M. J., "Methods and Indicators for Evaluating the Social Economy," in Bouchard, M. J. (ed.), *The Worth of the Social Economy. An International Perspective*, Brussels, Peter Lang/CIRIEC, 2009, pp. 19-34.

Bouchard, M. J., Ferraton, C. and Michaud, V., "First Steps of an Information System on the Social Economy: Qualifying the Organizations," *Estudios de Economía Aplicada*, Vol. 26, No. 1, April 2008, pp. 7-24.

Defourny, J. and Nyssens, M., "Approches européennes et américaines de l'entreprise sociale: une perspective européenne," *RECMA, Revue internationale de l'économie sociale*, No. 319, 2011, pp. 18-36.

Deshayes, G., *Logique de la coopération et gestion des coopératives agricoles*, Paris, Skippers, 1988.

ESSEC, *Guide du retour social sur investissement (SROI)*, Les Cahiers de l'Institut de l'Innovation et de l'Entrepreneuriat Social, Cergy Pontoise, 2011. (Translation and adaptation to French of Nicholls, J., Lawlor, E., Neitzert, E. and Goodspeed, T., *A guide to social return on investment*, Office of the Third Sector, The Cabinet Office, London, 2009).

European Commission, International Monetary Fund, Organisation for Economic Co-operation and Development, United Nations, World Bank, *System of National Accounts 2008*, New York, 2009.

Gadrey, J., "Les bénéfices collectifs des activités de l'économie sociale et solidaire: une proposition de typologie, et une réflexion sur le concept d'externalités," *Working Paper*, Université de Lille 1, 2002.

Handy, F., "Review of *"The Price of Virtue"* (2001)," *Nonprofit and Voluntary Sector Quarterly*, Vol. 32, No. 3, Sept. 2003, pp. 481-485.

Hausman, J. A. (ed.), *Contingent Valuation: A Critical Assessment*, North Holland, Amsterdam, 1993.

Hausman, J. A., "Contingent Valuation: From Dubious to Hopeless," *Journal of Economic Perspectives*, Vol. 26, No. 4, fall 2012, pp. 43-56.

Institut des Comptes Nationaux, *Le compte satellite des institutions sans but lucratif 2009-2010*, Banque Nationale de Belgique, Brussels, 2012.

Kahneman, D. and Knetsch, J. L., "Valuing Public Goods: The Purchase of Moral Satisfaction," *Journal of Environmental Economics and Management*, No. 22, Elsevier, Amsterdam, 1992, pp. 57-70.

Kaminski, P., *Les associations en France et leur contribution au PIB. Le compte satellite des institutions sans but lucratif en France*, Nanterre, ADDES, February 2006.

Mertens S., *Vers un compte satellite des institutions sans but lucratif en Belgique*, Thèse de doctorat en sciences économiques, Université de Liège, 2002.

Mertens, S. and Marée, M., "Les contours de l'entreprise sociale," in Mertens, S. (ed.), *La gestion des entreprises sociales*, Edipro, Liège, 2010.

Mertens, S. and Marée, M., "The Limits of the Economic Value in Measuring the Global Performance of Social Innovation," in Nicholls, A. and Murdock, A. (eds.), *Social Innovation: Blurring Boundaries to Reconfigure Markets*, Basingstoke, Palgrave MacMillan, 2012.

Mertens, S. and Marée, M., "La 'performance' de l'entreprise sociale: définition et techniques de mesure," *Revue Internationale PME*, Vol. 25, No. 3-4, 2013, pp. 91-122.

Mouchamps, H., "Weighing Elephants with Kitchen Scales. The Relevance of Traditional Performance Measurement Tools for Social Enterprises," *International Journal of Productivity and Performance Management*, Vol. 63, No. 6, 2014, pp. 727-745.

Olson, M., *Logic of Collective Action: Public Goods and the Theory of Groups*, Harvard, Harvard University Press, 1965.

Perrot, P., "Définition et mesure de la 'valeur ajoutée sociale' dans les associations," in *RECMA, Revue internationale de l'économie sociale*, Vol. 301, 2006, pp. 42-60.

Statistique Canada, *Compte satellite des institutions sans but lucratif et du bénévolat 2008*, December 2009.

United Nations, *Handbook on Nonprofit Institutions in the System of National Accounts*, New York, United Nations Publications, 2003.

PART II

WHAT CAN BE LEARNED FROM SPECIFIC STUDIES

The Construction of Social and Solidarity Economy Statistics in France

A Progressive Mobilization of Very Diverse Actors

Danièle Demoustier

Economist at Sciences Po Grenoble, France

with contributions from
Élisa Braley

Former head of the Observatoire national de l'Économie Sociale et Solidaire, France

Thomas Guérin

Project coordinator, Chambre Régionale de l'Économie Sociale et Solidaire Provence-Alpes-Côte d'Azur, France

Daniel Rault

Former technical advisor, Délégation interministérielle à l'innovation, à l'expérimentation sociale et à l'économie sociale, France

Introduction

The history of statistics on the social and solidarity economy (SSE) is, as with the history of general statistics in France (Desrosières, 2000), a social construction that reflects the mobilization as well as the representations and resistances of various social and institutional actors. SSE statistics in France took shape outside of an institutional environment with the recognition of the social economy by national actors and governments in the 1980s. Although INSEE[1] is a key actor in this process, statistics on

---

[1]   The National Institute of Statistics and Economic Studies (INSEE) is a branch of the Ministry of Economy and Finance in France. It collects, produces, analyzes and

the SSE are still produced by multiple private actors. Moreover, despite a growing consensus on the weight of the SSE in the overall economy, even leading to the appointment of a Minister of the SSE and to the adoption of a legislation,[2] and despite the fact that the figures and the longitudinal monitoring are taking shape, many questions remain regarding the scope, the categories and the indicators.

Overall, the institutionalization of SSE statistics is still a work in progress. We attribute this delay to the origin and nature of the issues under debate and to the still considerable diversity of the actors.

This chapter first recalls the steppingstones leading to the production of statistics on the social and solidarity economy in France, starting from the 1980s. We then expose the different stakeholders that took part in the process and their viewpoints on the matter. In conclusion, we see the importance key social actors played in ameliorating the quality of the data produced on social and solidarity economy in France, contributing to its better public and institutional recognition.

1. A recent construction comprised of successive adjustments and improvements

The history of social and solidarity economy statistics evolved in three phases, each spanning over one decade: the 1980s, characterized by movement and appeals; the 1990s, when the government gradually became involved, and the 2000s, during which the observatories were built.

The 1980s – a time of appeals and high ambitions

Once the shared identity of the actors was recognized (notably through the *Charte de l'économie sociale* adopted in 1980),[3] and once recognition was obtained from the government (through the creation of the

---

disseminates information on the French economy and society; it coordinates statistical monitoring and contributes to building an international statistical space. As of 2011, it also oversees and supervises the higher education and research activities of the Groupe des écoles nationales d'économie et statistique (GENES).

[2] In France, ministerial reshuffling and political changes starting in the 1980s have led to the creation, in 2012, of a ministry dedicated specifically to the social economy, which was in 2014 incorporated into the new Secrétariat d'État au Commerce, à l'Artisanat, à la Consommation et à l'Économie sociale et solidaire. The same year, France adopted a Law on social and solidarity economy.

[3] This charter states, in seven articles, the values, principles and characteristics uniting the organizations that recognize each other in the context of the social economy. It was updated in 1995.

Délégation interministérielle à l'économie sociale, DIES,[4] in 1981), the question of internal knowledge about the social economy was raised. The Association pour le développement de la documentation sur l'économie sociale (ADDES), which had been created by the initiative of Crédit Coopératif, a bank specifically dedicated to serving the social economy, backed by a number of prominent researchers and statisticians, called for the construction of a reliable statistical system about what had until then been regarded as a nondescript cluster of very disparate organizations.

Through an annual conference organized by its technical councils, ADDES (ADDES, 1983-2013) sought to highlight existing studies and to appeal to the government. Its first achievement was the construction of a satellite account capable of identifying and presenting the various contributions of the social economy into the various income, output and employment streams. For many years, this project benefitted from the input of expert statisticians in national accounting from INSEE, which even assigned an executive from DIES to assist in the building of the account.

The first results were based primarily on registers (from federations, SIRENE[5] business files) and secondarily from the first data aggregated by INSEE (in terms of businesses, establishments and employment) from social data.[6] In accordance with usual practice, INSEE wanted the ministry responsible for the social and solidarity economy sector to support the satellite account beyond the start-up phase. This work was then entrusted to a statistician appointed specifically for this purpose by DIES.

At the same time, the economist Claude Vienney, then teaching social economy at the continuing education division of the University Paris I Panthéon Sorbonne, sought to establish an overview of the emerging university programs on the social economy. For this, he created BTI DOPERES (Banque de travail interuniversitaire pour la documentation des programmes d'enseignement et de recherche sur l'économie sociale), an interuniversity databank operated by a small group of academics

---

[4] A commission attached to the office of the Prime Minister to improve relations between the state and the social economy and to promote the social economy. It was coordinated and overseen by the ministries of, in that order, Affaires Sociales, Jeunesse et Sport, and Emploi. In 2010, it was incorporated into the Direction de la Cohésion Sociale.

[5] National identification system and register of companies and their establishments. Managed by INSEE, Sirène assigns a SIREN code to businesses, organizations and associations and a SIRET code to their establishments and facilities.

[6] These data were derived from the annual declarations of social data which all employers of salaried staff must submit annually under the French social security system.

who compiled information about the various components of the social economy.[7]

This was the infancy stage, characterized by questioning and, as we just observed, by high ambitions.

The 1990s – the government gradually gets involved

In the late 1980s, the Secrétariat d'État in charge of the social economy agreed with INSEE to respond to the appeals of the social economy sector. At that occasion, and following assessments of the previous results, needs and objectives, DIES (through INSEE) assigned yet another technical adviser to ADDES.

However, the audit report submitted to an administrator of INSEE in early 1990 expressed skepticism about the social economy. Indeed, it recommended drawing a clear line between cooperatives and mutuals (easily identifiable as companies) and associations (a vague category). The report also maintained that nothing had been done in the way of establishing statistical data for associations; that the development of a satellite account for the entire social economy was no longer a priority; and that more emphasis should be placed on the collection of statistical data of greater value to the stakeholders and the recognition of their activities.

In this perspective, the main groundwork was laid in the early 1990s by the main types of legal status, based on data from the federations, and presented in four issues of the *Revue des Etudes Coopératives*[8] (insurance mutuals, manufacturing workers' cooperatives, health mutuals, agricultural cooperations) (RECMA, 1991). A progress report of the national statistics was submitted by the DIES consultant at each annual meeting of ADDES; nevertheless, the actors still expressed frustration about the slow progress made by the public authorities.

Meanwhile, study and research groups conducted regional studies, which were considered important by the DIES consultant for strengthening ties with the regional directorates of INSEE.

Throughout the 1990s, the activities of associations gained increasing importance, to the extent that they could no longer be overlooked by statistics. At the end of the decade, at the request of the Prime Minister,

---

[7]   This ambitious project, no doubt ahead of its time, had only a brief existence despite the support of DIES.

[8]   The *Revue des Etudes Coopératives* was founded by Charles Gide and Bernard Lavergne in 1921. It has since been extended to include mutual societies and associations and is today a reference journal in the francophone world (*RECMA, Revue internationale de l'économie sociale*).

a report was issued by the Commerce-Services section of the Conseil national de l'information statistique services[9] (CNIS, 1998, 2010). It found that associations were being monitored only through questionnaires and surveys, with INSEE conducting surveys among households and Viviane Tchernonog among associations (Tchernonog, 2007, 2013). Calling for a closer monitoring of enterprises incorporated as associations, the report laid the groundwork for an annual survey program for sectors with a strong contribution coming from associations. The latter, mostly non-commercial in nature, had not been covered by the annual survey program of INSEE. On that basis, the report recommended that considerable investments be made by the national statistics agency, albeit without imposing an economic vision on the associations. The report also emphasized the importance of a broader monitoring of the evolution of associative practices.

Three recommendations made in the report were implemented as a priority: filtering SIRENE for associations with salaried employees; the customized exploitation of the DADS[10] in order to develop an annual table of jobs; and the matching up of the turnover tax declarations files sales with those submitted to SIRENE. This generated the necessary data for the following years about the weight of associations in the social economy. In accordance with DIES, the division of national jurisdiction at INSEE's regional office in Midi-Pyrénées developed a "kit" of 24 statistical tables about the social and solidarity economy (after the usage of the term social economy in the 1970s, "social and solidarity economy" (SSE) terminology was adopted starting in 1999), which guaranteed a succinct yet annual and regular statistical production derived from administrative sources that had become more reliable in this field.

Overall, the 1990s phase comprised the first wave of realizations, characterized by regional activity (researchers and regional directorates of INSEE) supported by DIES at the national level. But the results were that: the satellite account project was put on hold; the SSE was equated as being a set of companies with a specific status; and, with data remaining cursory, no headway was made in treating associations any differently.

The 2000s – the construction of observatories

Starting in the early 2000s, a major research program was implemented by DIES, with the assistance of the Mission recherche du ministère des

---

9   Place of consultation between producers and users of public statistics created in 1984 to discuss new needs and to participate in the development of the national statistics program.

10 The annual declarations of social data, which employers are required to file annually, consist of the number of staff and their gross pay, i.e., the sums from which social contributions are calculated.

Affaires Sociales (MIRe), to meet the guidelines proposed by CNIS (including recommendation 13: "bolster the statistical investment through appropriate studies"). At the same time, regional chambers of the social (and solidarity) economy took shape, hiring employees and generating a strong need to assess the weight of the SSE in the regions. Regional studies were conducted in a somewhat uncoordinated fashion and the first regional observatories were created (e.g., Provence-Alpes-Côte d'Azur, Aquitaine) in 2007. DIES, INSEE and the Conseil national des chambres régionales de l'économie sociale et solidaire (CNCRES)[11] then became aware of the wide range of measurement methods in practice, and that this rendered impossible the aggregation of regional measurements of the SSE as well as a comparison with other economic actors. These three actors then met and established a perimeter and a harmonized methodology that led in 2008 to the development of a permanent observation and measuring system of the SSE in France and the regions (the "kit ESS" or "Investissement E12B" realized by INSEE Midi-Pyrénées) and the first national publication by the national INSEE.[12]

As places of production and sharing of work on the SSE, the observatories apply a standardized, scientific and consistent methodology between the territories. This allows them to perform their functions with regard to producing studies, monitoring, providing decision aid and prospecting from which the SSE actors as well as the government benefit when determining their territorial policies. The primary source of data for the observatories is INSEE (DADS, business directories, local data, etc.), which they supplement with their own surveys as well as data from other observatories, providers of studies and statistics, and networks of actors and public institutions – together comprising some hundred sources in France.

The observatories have steering committees as well as scientific committees to ensure the quality of productions and a critical and multidisciplinary approach, both for the definition of specific indicators of the dynamism of SSE organizations and enterprises and to improve the monitoring. Since 2009, the observatories produce regional and infra-regional summary statistics, whereby the national council of CRESS is in charge of the national monitoring and the production, with the help of researchers, of the *Atlas commenté de l'ESS*. The latter presents statistical

---

[11] Created in 2004 by the Chambres régionales de l'économie sociale (et solidaire) (CRESS), the national council of CRESS is a place for dialogue and resource sharing among its members. It manages a network of 15 regional observatories and publishes every two years a *Panorama synthétique de l'ESS* and an *Atlas* with commentaries by researchers.

[12] Guillaume Gaudron (2009), "L'économie sociale emploie un salarié sur 10 en 2006," *INSEE Première*, No. 1224, February 2009.

tables, geographical distribution maps and commentaries and analyses about a number of developments and issues (CNCRES; 2009, 2012, 2014).

Quantitative presentation has seen tremendous progress over the years. It allows for meaningful comparisons over time and with other types of businesses (non-SSE private firms and the public sector). In parallel to the development of statistical series, INSEE has recommended making analyses that complement previous categories, such as the analysis of salary differentials and estimates of the overall added value – which again raises the question of the internal structure (especially that of the integration or not of the economic activities of subsidiaries with those of members, while the competing groups show "consolidated" data). In addition, the refinement of territorial analyses (by region and then by department and employment area) raises new questions: Does the same weight translate into the same reality? Does a high weight necessarily indicate a strong influence?

Furthermore, the 2008 crisis, which impacted not only market activities but also public funding, called for a cyclical monitoring that is more responsive than the biannual relaying of information to INSEE. Through a direct contact, the association Recherches et solidarités (2009, 2013) has been able to promptly publish current economic data on a slightly different statistical field (employers contributing to URSSAF[13] and the Mutualité sociale agricole).

To enhance the mechanism, the *Loi sur l'économie sociale et solidaire* (SSE Act), passed in 2014, recognizes the coordinating role of CNCRES in data collection and provides for, in Article 8, the "statistical monitoring," mobilizing the public banking system (Banque de France, Banque publique d'investissement) to provide data on economic activity and origin of revenue. Still, the implementation of the SSE Act does not provide for the full integration of the SSE into the general statistical analysis, nor is it an exercise in which economic data is complemented with social data concerning the specificities of the SSE such as its volunteer work or its membership base.

However, a subsequent step sees for a more qualitative analysis of solidarity models underlying economic models. The analysis examines

---

[13] URSSAF (Unions de Recouvrement des Cotisations de Sécurité Sociale et d'Allocations Familiales) are organizations mandated by government to collect social security and family benefit contributions from companies. These contributions finance France's main social security system and related programs and institutions (unemployment insurance, transport authorities, national housing funds, pension funds, universal health coverage, and supplementary pension schemes) (Adapted from Wikipedia, accessed July 2, 2014).

impacts such as avoided costs or generated revenue in order to better specify the performance and efficiency indicators of SSE firms and to assess their "social utility."

2. Stakeholder analysis: a compromise on the path to stabilization?

This tentative construction of SSE statistics demonstrates the instability of the referential, given the diversity of the types of actors involved (in terms of geography, expertise, legitimacy, etc.) and their views, concerns and goals. However, the SSE Act attempts to stabilize the scope and role of not only the actors but also of the statistical production, even if at present the SSE can only resort to a partial monitoring system.

The diversity of social representations of the SSE has not disappeared, as demonstrated by the variety of actors and their objectives. The scope, and its place and role, continue to be a subject of debate. In that context, the representatives of national federations and associations seek to highlight the value of their membership base, be it towards competitors in the market (for the cooperative sector) or towards public authorities (for associations). These national representatives have witnessed changes, prompted mainly by external reasons. First the social economy (national) extended to the solidarity economy (local) in the 1990s, and then, in the 2000s, to social entrepreneurship (in the form of commercial businesses with a social purpose). Conversely, the national associations generally argue for a broad definition of their own field: cooperatives by including their subsidiaries and members (which raised employment in the sector from 300,000 to 1 million); associations by questioning the reduced economic activity to that of employers (which reduced the number of associations by at least 200,000 for the 1 million associations estimated to be active). They highlight either the economic contribution in terms of turnover and employment, or in terms of volunteering and social cohesion, or the trickle-down economic impacts of organizations that hire few employees (such as the Coopératives d'utilisation de matériel agricole, the Associations pour le maintien d'une agriculture paysanne or the Structures d'insertion par l'activité économique).

The researchers, for their part, do not all share the same vision of the SSE sector. This is despite the fact that they are less and less isolated from one another,[14] as they are linked to laboratories and organizations, and despite exchanges within the Réseau inter-universitaire de l'ESS (RIUESS).

---

[14] In relation to the first to invest the field, such as of P. Kaminsky, J.P. Le Bihan and C. Vienney.

This network also convenes annual conferences, bringing together the non-profit sector, cooperative economy, solidarity economy, social economy and social entrepreneurship. These concepts intersect both in the empirical field and in theory, and are complemented by a variety of study topics such as activity, public, mode of governance, employment, impact, regulation and territorial development. In this way, researchers are advancing in terms of the refinement of the analysis but this division of the scientific labor offers little in terms of the overall vision of the social an solidarity economy.

At the regional level, however, the CRESS combined the approach of federations, partners and researchers (having set up a scientific committee) by mobilizing public and private data and by conducting their own surveys. Through its coordination function, CNCRESS takes responsibility for the broader matter, whereby its mobilization of large bodies of data is more a juxtaposition than an integration of data (see *Atlas*). Other intermediate organizations (such as ADDES or Recherches et solidarités) reinforce this approach either through more qualitative analyses or through more current statistics.

On the part of government institutions (DIES then eventually the Ministry in charge of social economy, or INSEE), progress takes place in small steps: from the assignment of a resource person to the mobilization of a regional directorate, investment has become more institutional. At the national level, INSEE has been publishing statistical series from the "kit" since 2005 on companies, establishments and the jobs of the SSE. These institutional statistics focus on the economic activity and the generated employment, objectives that are reinforced by the obligation for statistical monitoring under the SSE Act. Business logic here prevails at the expense of that of collective organization. The national survey project to be conducted by INSEE on all associations in 2014 could in this sense poorly align with this SSE statistic.

This diversity has a number of uncertainties.

The first uncertainty concerns the perimeter to study, namely that of enterprises incorporated as associations, cooperative groups (subsidiaries and members), and companies that do not have a cooperative, mutual or association legal status (social enterprises, social entrepreneurship) despite the SSE Act, which makes specific mention of "social and solidarity economy associations" and which defines which companies without a status can claim to be part of the SSE.

The second uncertainty applies to the relevant categories, such as enterprise/group, activity/action, employment/engagement, product/process, or relationship. This makes it difficult, in the monitoring of jobs and work, to integrate not only assisted contracts but also the intermediate statuses, such as volunteer work and salaried entrepreneurship. Similarly, the monitoring of activities does not take into account the diversity and the crossovers between various activities such as popular education or social integration by economic activity, where it is difficult to distinguish the principal activity (training or building, for example).

Another questioning addresses the nature of the data to be collected: economic (market, non-market, non-monetary) and social (engagement, membership base, solidarity, inclusion, employment); quantitative and qualitative; and national and local. This leads to juxtapose data without being able to fully integrate them, as this would require a modification of the categories of the national statistics, which was not designed for the SSE and which calls for a mingling of economic and sociological data. The methods to implement also raise questions: Simple aggregations? Comparisons with "the other classical economics"? or Focus on the specificities which new indicators would impose?

Finally, an overall uncertainty exists about the use of these statistics: Standardization or differentiation? Common access rights or construction of specific policies? Contribution to growth or emergence of new development models?

Conclusion

In the absence of a dominant actor playing an active role, the multiplicity of actors involved has been a motor more than a brake in the improvement of SSE statistics, despite differences in views and goals. The aspect concerning quantitative statistics was nevertheless an important tool to promote recognition, in the sense of "what can be counted, counts."

However, this trial-and-error approach and this mobilization of sources and various producers led more to a juxtaposition than to a full integration, despite the coordination by CNCRES. Any real advances therefore appear uneven, slow and partial. The legitimacy of the data is not always a given, coming as they do from different sources (e.g., figures produced by the federations; national surveys on businesses; surveys on associations; surveys on businesses; figures on volunteer work; surveys about associations and household surveys).

Overall, data is being gathered more on the basis of what is available than of what is necessary. However, the diversity of approaches – underlying the data – allows for a broader collection. It creates an incentive between data producers by the questioning and the mutual adjustments. Thus, the non-rigid system is still evolving. The original actors (mostly from the political realm and scientists) have given way to more technical organizations that structure the data in a normative framework.

Thus, ADDES, which initiated the construction of the statistical system, today advocates the adoption of more qualitative and strategic analyses, while INSEE produces standardized statistical series. However, achieving a true institutionalization of SSE statistics raises the question of the adaptability of the national statistics system. The latter is by definition normative, which is necessary for comparisons across space and time, but still restrictive concerning the specificities of the SSE.

References

ADDES, *Actes des colloques*, 1983-2013, http://www.addes.asso.fr/.

Chadeau, A., "Peut-on croire en l'économie sociale?," in *RECMA*, No. 38, 1991.

CNCRES, Observatoire national de l'ESS, *Panorama national de l'ESS en France et dans les régions*, 2008, 2010 and 2012.

CNCRES, ARF, CDC, Chorum, *Atlas de l'économie sociale et solidaire en France*, 2009.

CNCRES, Observatoire national de l'ESS, *Atlas commenté de l'économie sociale et solidaire*, 2012 and 2014.

CNIS, *Rapport de la mission "Associations régies par la loi de 1901,"* No. 44, November 1998.

CNIS, *Rapport du groupe de travail "Connaissance des associations,"* No. 122, December 2010.

Desrosières, A., *La politique des grands nombres, histoire de la raison statistique*, Paris, La Découverte, 2000.

Draperi, J.-F. (dir.), *L'année de l'économie sociale et solidaire*, Paris, Presses de l'économie sociale, Dunod, 2010.

Recherches & Solidarités, *Economie sociale: Bilan de l'emploi en 2010*, (and 2011, 2012, 2013) http://recherches-solidarites.org/etudes-thematiques/economie-sociale/.

RECMA, Avant-propos pour une nouvelle chronique; les mutuelles d'assurance; les SCOP; la coopération agricole; la mutualité, Nos. 37 to 40, 1991.

Tchernonog, V., *Le paysage associatif français, mesures et évolutions*, Paris Juris Association-Dalloz, 2007, 2013.

The Production of Statistics for the Social Economy in Belgium

A Focus on Mutuals and Cooperatives

Fabienne FECHER

Full professor, Faculty of Social and Human Sciences,
Université de Liège, Belgium

Wafa BEN SEDRINE-LEJEUNE

Université de Liège, Belgium

1. Introduction

1.1. From a statistical need ...

The lack of visibility of cooperatives and mutuals in present-day societies contrast with the increasing importance of those organizations, which are firmly established in every sector and branch of economic activity. Given the activities that they carry out and the goals that they pursue, the cooperatives and mutual societies are two essential components of the social economy. Through their mission of combining objectives that are economic as well as social, they represent readily identifiable forms of enterprises in the social economy. For those organizations, the creation of an economic added value is more of a means of achieving their objectives and values rather than the be-all and end-all of their activities.

The EU has some 246,000 cooperative enterprises with some 144 million citizen members, employing some 4.8 million persons. Nearly 120 million Europeans are covered by a health mutual and the mutuals hold a significant share of the life insurance and non-life insurance markets. Mutual societies represent 25% of the European insurance market and 70% of the total number of insurance companies in Europe.

In Belgium, the social situation during the 1970s gave rise to new initiatives in the social economy to hammer out solutions to problems in the field of employment and the environment. Today, new cooperatives have come into being to provide mutual solutions to the economic and social problems of households, companies and social organizations. We thus see recent initiatives in fields such as sustainable energy production (for example, wind energy), alternative finance, integration through work and social and health care.

In the historico-institutional context of Belgium the mutual societies have existed for a long time. Even in the 19th century they were giving their members allowances in case of illness or death in return for membership fees. With the post-war years of full employment they gradually became integrated in the public social security system. The mutual societies are now closely linked to the management of health insurance. Beside those health mutuals, Belgium has insurance mutuals that set themselves apart from 'classical' insurers, by their primary purpose, which is to develop a relation of lasting confidence with their members with a view to covering their insurance needs and not to make profit in order to remunerate contributors of capital.

Barea and Monzón (2006) explained this paradoxical institutional invisibility of those enterprises of the social economy by two main reasons.

Firstly, the lack of a clear definition of the concept and scope of the social economy, of the shared characteristics of the different classes of companies and organisations that are part of it and the specific traits that enable them to be distinguished from the rest of the organisations that move in the economic system makes it difficult to delimit the field of study accurately and to identify institutional units with shared characteristics and homogeneous economic behaviour at the international level irrespective of legal and administrative criteria, which are very diverse and mutually contradictory from one country to another.

Secondly, the system of national accounting, that is the United Nations' 1993 System of National Accounts – NAS 1993 and the European System of National Accounts – ESA 95 (Eurostat, 1996), has developed tools for collecting the major economic aggregates in a mixed economy context with a strong private capitalist sector and a complementary public sector. Logically, in a national accounts system which revolves around such a bipolar institutional situation there is little room for a third pole which is neither public nor capitalist, while the capitalist pole can be identified with practically the whole private sector. As a consequence the companies and organisations of the social economy disappear into the different institutional sectors established by the national accounts systems. Thus, cooperatives are considered as being commercial companies that share out their profits among their members and the mutual societies are treated as being financial institutions that belong to the sector of enterprise.

To improve the identification and quantification of the enterprises of the social economy, several efforts were undertaken at the national level and by the international institutions (for instance the European Economic and Social Committee) during the 1980s. At the Belgian level, an observatory system, called ConcertES, was set up in the Walloon Region and the Flemish Government called for the development of an observatory system for social integration companies. At the European level, the efforts culminated in 1997 in the publication by Eurostat of a report on the Cooperative, Mutualist and Associative Sector in the European Union (Eurostat, 1997).

1.2. … To satellite accounts

For completing and improving the statistics collected on social economy and considering the inability of the central framework of national accounts to measure certain specific areas of economic and social life, the satellite accounts show themselves to be a useful instrument.

"A satellite account is an evolutionary framework that brings together the data for a field of economic or social concern, offering more detailed and flexible information than that provided by the central framework of the national accounts to which it is linked, which constitutes its frame of reference" (Archambault, 2003). So, the satellite account has the vocation of prolonging and completing the conceptual framework of the national accounts. It constitutes a means of structuring the quantitative information relating to a particular area of study (here, the field of enterprises of the social economy) by offering a coherent system of statistical information that may be used for the purposes of macroeconomic analysis by the managers in that field, public decision-makers, experts or any other interested party.

Convinced by the interest and the power of that tool, the United Nations has sponsored the development of a manual that establishes a satellite account for Nonprofit Institutions (NPIs), a group that covers both non-profit institutions serving households and all the private non-profit entities that are dispersed among the other institutional sectors. This group excludes cooperatives and mutual societies from the non-profit sector. The Handbook (United Nations, 2003) established the first officially sanctioned procedure for capturing the work of non-profit organizations in national economic statistics, totally structured and consistent with the national accountancy system ESA 95. Belgium was one of the three leading countries (with Italy and Australia) to publish, in early 2004, the first satellite account of non-profit-making institutions. Since the publication of the handbook in 2003, 16 countries have produced NPIs satellite accounts for at least one year.[1] The implementation of the handbook is carried out by government statistical agencies. The main benefit of this approach is that the NPIs satellite accounts have the status of official statistics.

---

[1]   See the chapter by Salamon, L.M., Sokolowski S.W. and Haddock, M. in this book.

In order to prolong and complement this Handbook, the European Commission entrusted CIRIEC[2] with the task of writing a manual on the satellite accounts of cooperatives and mutual societies, which was published in 2006 (Barea and Monzón, 2006). Its first purpose was to establish a rigorous conceptual demarcation of the social economy companies to be studied in the satellite accounts. The proposed definition attained a wide consensus, both among the most prominent organizations that represent the social economy in Europe and in the sphere of the specialist literature in this field of economics. The second aim of the manual was to establish the guidelines and a methodology that will allow the satellite accounts for the cooperatives and mutual societies to be drawn up in accordance with the central national accounting framework established by the ESA 95. As a consequence, the satellite accounts for cooperatives and mutual societies are complementary of the NPIs' satellite accounts.

Funded by the Enterprise & Industry Directorate-General of the European Union, a large project was launched in 2010 with the aim of producing first satellite accounts for cooperatives and mutual societies by the Member States of the European Union. This project served the European Commission Competitiveness and Innovation Framework Programme and the Programme for Innovation and Spirit of Enterprise and arose from the observed fact of the lack of institutional visibility of the enterprises of the social economy. Five countries were selected: Belgium, Bulgaria, the Republic of Macedonia, Serbia, and Spain.[3]

The present chapter summarizes the Belgian report, SATACBEL, (Ben Sedrine *et al.*, 2011) on the establishment of keys indicators for satellite accounts in Belgium covering cooperatives and mutual societies for the year 2007. Several macroeconomic aggregates are assessed and examined with regard to the place that cooperatives and mutual companies occupy within the Belgian economy. To achieve this goal a close cooperation was necessary with the Belgian National Accounts Institute (NAI) within the Belgian National Bank (BNB).[4]

Thanks to this work, political decision-makers and interested parties have at their disposal strict methods and primary information and indicators allowing evaluation of this particular field of the social economy

---

[2]   International Centre of Research and Information on the Public, Social and Cooperative Economy.

[3]   An international seminar to exchange experiences and good practices and to discuss about the results obtained in the five countries took place in Madrid in July 2011.

[4]   We must offer our special thanks to Catherine Rigo and Marie Vander Donckt, who so ably helped and supported us throughout this research. However, the scientific content of the report remains the sole responsibility of the authors.

sector. Because the national contexts and realities behind cooperatives and mutual societies are very different from a country to the other, the results obtained should be compared internationally with caution.

Sections 2 and 3 define the populations under study on the basis of juridico-institutional criteria and behaviors (respect of several principles). The methodology is explained in Section 4. The results obtained are set out in Section 5 per group, sector and branch. The paper closes with a brief conclusion presented in Section 6.

2. Social economy in Belgium and the satellite account population

2.1. From the Belgian definition of social economy ...

In Belgium the most generally accepted meaning of social economy is used that is, *the social economy consists of economic activities producing goods or services, conducted by societies, mainly cooperatives and/or companies with social purpose, associations, mutual societies or foundations, whose ethic finds expression in all of the following principles:*

1. objective of service in the group or general interest, or for the members, rather than mere pursuit of profit;

2. autonomy of management;

3. democratic decision-making process;

4. primacy of people and labor over capital in the distribution of income.

In particular, this definition was hallowed by the Walloon Region's Decree[5] of 20 November 2008. It draws inspiration from the definition adopted by the Walloon Council for Social Economy (*Conseil Wallon de l'Économie Sociale*) in 1990.

Let it be noted that this definition is not exactly congruent with that which used to prevail in Flanders. Indeed, the concept of social economy in the North of Belgium traditionally covered only training and socio-professional integration projects. Nevertheless, a broader understanding of the SE has been proposed by the Flemish platform VOSEC (*Vlaams Overleg Sociale Economie*) that now approximates more closely the French conception applied here.

We would point out that this definition has been accepted at federal level in the context of agreements towards cooperation with the regions.

---

[5] Moniteur belge, 31 December 2008.

It also constitutes a reference framework that has inspired many countries and is totally congruent with the working definition of the cooperatives and mutual societies proposed by Barea and Monzón in their Manual (2006):

> The set of private, formally-organized enterprises with autonomy of decision and freedom of membership, created to meet their members' needs through the market by producing goods or providing services, insurance or finance where decision-making and any distribution of profits or surplus among the members are not directly linked to the capital or fees contributed by each member, each of whom has one vote.

Of course this definition applies to the part of the social economy which is made up of the market producers in the social economy, which are not covered by NPIs Handbook.

2.2. … To a definition of target population

The working definition proposed by Barea and Monzón (2006) remains the reference for the identification of enterprises to be taken into account for the construction of the satellite accounts. For the purposes of this project, three types of enterprise will be used for Belgium.

2.2.1. The cooperatives accredited by the National Cooperation Council

Under Belgian Law the cooperative society is governed by Commercial Law in the same way as any other commercial company. It may take the form either of the limited-liability cooperative society (*société coopérative à responsabilité limitée* – SCRL) or the unlimited-liability cooperative society (*société coopérative à responsabilité illimitée* – SCRI).

It is defined as being "made up of members the number of which and contributions from may vary" (*Commercial Law, Article 350*). This definition recognizes nothing of the cooperative spirit other than the variable nature of members and contributions.

The legislator has provided a system more flexible, less burdensome than that for other commercial companies; this soon attracted the attention of many entrepreneurs who saw in it, if anything, a way of eluding more restrictive rules without necessarily being moved by any cooperative ideal. These are known as the 'false' cooperatives.

Seeing the emergence of these cooperatives having chosen this status for no reason other than legal convenience, the legislator reacted, for example, by creating the National Cooperation Council (*Conseil National de Coopération* – NCC), providing the possibility for cooperatives respecting the principles of the social economy to set themselves apart as such by seeking accreditation from the NCC.

It was also to discourage certain abuses, giving it a suspect reputation, that the status of cooperative was revised during the 1990s (Law of 20 July 1991, revised in 1995). However, there are some who believe that this revision errs on the side of excessive caution, with a statute perceived as being overly restrictive.

It may be also noted that accreditation by the NCC allows the co-operative society to benefit from a favorable system – albeit of limited range – as regards taxation.

Finally it has to be said that accreditation by the NCC does not, however, allow a line of demarcation to be drawn between 'true' and 'false' cooperatives. In fact, while accreditation allows attestation of the fact that the cooperatives concerned fully adhere to the principles of the social economy, it does not allow the coverage of all the cooperative societies operating in that same spirit since,

(i) many cooperative societies chose not for NCC accreditation but for the transversal status of 'social purpose'

(ii) and we cannot exclude the existence of cooperative societies not having chosen NCC accreditation (nor even the 'social purpose') but still working in the spirit of the social economy by respecting the principles and ideal of cooperation.

2.2.2. The mutuals

❖ *The mutual health societies*

In Belgium, the mutual societies are closely linked to the management of health insurance and act as interface between the National Institute of Health and Invalidity Insurance and the citizen. This part of their activities, falling within the obligatory area, may not therefore be considered as being of the social economy and must consequently be excluded from our field of study.

However, they also develop, and more autonomously, other activities in connection with free or complementary insurance and various social services. This strand of their activities is included in our field of study.

❖ *The mutual insurance societies*

Under Belgian Law a mutual health insurer may take the legal form of the *caisse commune* or the *société d'assurance mutuelle*. Like any other insurance company it is subject to accreditation from and control by the Banking, Finance and Insurance Commission (*Commission Bancaire Financière et des Assurances*).

A mutual insurance society is defined as an insurance company that is the collective property of its members and acts in the interests of its members.

The insurance mutuals thus set themselves apart from 'classical' insurers, first of all by their primary purpose, which is to develop a relation of lasting confidence with their members with a view to covering their insurance needs (and not to make profit in order to remunerate contributors of capital). The mutual insurance society does not have any shareholders to remunerate.

The characteristic of 'collective property of its members' derives from the fact that the member has the dual aspect of insured individual and collective insurer. This specificity finds expression in, among other things, an involvement of the members in company policy.

The insurance mutuals are therefore squarely situated in the field of the social economy and perfectly match the criteria for inclusion in our study population.

2.2.3. The companies that pursue a social purpose

The company with a social purpose (CSP) was born of the will to bridge a certain legal gap allowing a combination of social purpose and commercial activity in a context of redeployment of the social economy. It was necessary, D'Hulstère and Pollénus point out "for this resurgent social economy to find a middle path between the status of commercial company (including that of the cooperative society), that supposes a profit motive, and the status of non-profit-making institution not allowed to pursue a commercial activity" (D'Hulstère and Pollénus, 2008, p. 26).

The status of CSP was introduced into Belgian law by the Law of 13 April 1995. The CSP is a commercial company governed by Commercial Law (*Articles 661 to 669*).

The CSP is not a new form of company but, rather, a variant that may be adopted by different commercial companies already in existence. In other words, it has no life of its own but grafts itself on to one or another of the classical legal forms for which provision is made under Commercial Law, namely the general partnership; the limited partnership with share capital; the limited-liability company; the limited-liability or unlimited-liability cooperative society; the public limited-liability company; the simple limited partnership; the economic interest grouping.

However, to claim the label 'social purpose', the Articles of Association of a company must embody the nine additional points[6] as provided for by Article 661, § 1 of the Code of Commercial Law (see Annex 1). So the

---

[6] The nine principles of the CSP are completely in harmony with the spirit of the four principles inherent in the social economy and also honor the principles of the cooperative ideal.

great majority of companies to have adopted the status of social purpose had also attended the birth of the cooperative societies.

3. Demography of population and branches of activity

3.1. Demography of the population

It is on the basis of a juridico-institutional referencing that this step will be accomplished before culminating in the actual composition of our population after the cleaning of the data (empty cells, repeats, reclassifications, …). Our study thus ultimately covers a population of:

- 461 cooperatives accredited by the National Cooperation Council (*group 1: 'NCC' coop*);
- 421 companies with social purpose (*group 2: 'CSPs'*);
- 18 insurance mutuals and community chests (*group 3.1: IM & CC*) and
- 5 national unions of health mutuals (including all the mutualist entities that they cover, but excluding their activities classified as obligatory insurance) (*group 3.2*).

It should be pointed out that, at this stage, the cooperatives accredited by the NCC (and therefore entering our field of inclusion) represent only a minute fraction of companies legally constituted in the legal form of the cooperative (that is, only slightly more than 1% of the 40,000 Belgian cooperatives). Now, to recall, although the NCC accreditation allows *a priori* attestation that those companies adhere to the principles of the social economy, we cannot exclude the existence of cooperatives not having chosen in favor of 'NCC'-accreditation, but working all the same in the spirit of the social economy. Indeed, as Van Opstal reminds us, "it would be too easy to conclude that the 39,500 others had nothing to do with cooperative ways of doing business. To be sure, we have found that many cooperatives without accreditation present marked similarities as regards activity, membership profile and functioning with certain of the accredited cooperatives" (Van Opstal *et al.*, 2008, p. 152).

3.2. Branches of activity in the population

The study of the demography of the population allowed the segmentation by activity branches in accordance with the central framework of the national accounts. The activity branches based on the 2-digit NACE divisions[7] present in our three groups of population under review (*'NCC*

---

[7] The complete 2-digit NACE A60 classification is given in Annex 2.

cooperatives', CSPs, mutuals) are given in Table 3/1. Some of our results will be presented per branch in section 5.1.3.

Table 3/1. *Activity branches NACE reference A60*
by group of population

Activity branches present in the population		
Cooperatives **(group 1)**	**CSPs** **(group 2)**	**Mutuals** **(group 3)**
01; 02	01; 02	
15; 22; 28; 29; 36; 40	15; 18; 20; 22; 26; 28; 29; 33; 36; 41	
	37	
45	45	
50; 51; 52	50; 51; 52	
55	55	
60; 61	60; 61; 63	
65; 66; 67	67	66
70; 71	70; 71	
72; 73	64[8]; 72; 73	
74	74	
	80	
85	85	
91; 92; 93	91; 92; 93	

4. Construction of aggregates – Methodology

4.1. Reference framework

Our satellite account being founded on the conceptual framework proposed by Barea and Monzón (2006), the main methodological reference remains, first and foremost, the framework defined in the *European System of National and Regional Accounts* – ESA 95.

To recall, the ESA 95 is the European version of the *National Accounting System* (NAS 1993) established under the combined aegis of the United Nations, the International Monetary Fund, the World Bank, the Organization for Economic Cooperation and Development and the European Community.

The Belgian National Accounts Institute has produced an operational document[9] that describes the methodology applied at Belgian level to de-

---

[8] Our sample counts 1 single unit from the branch NACE 64 (telecommunication) and active in the computer branch; this was integrated to the branch NACE 72-73 (computer and R&D services).

[9] Downloadable document on the following URL: http://www.nbb.be/doc/DQ/ F_method/M_Inventaire_SEC1995_FR_def.pdf, consulted in March 2010.

termine the Gross Domestic Product (GDP) at current prices. It is, as it were, a practical guide for the application in Belgium of the ESA 95, account being taken of the relevant institutional, legislative and accounting context in Belgium. We derive the essentials of our lines, choices and methodological conventions from this manual, generally referred to by the experts at the Belgian National Bank as the '*Inventaire*'.

4.2. Aggregate calculation methods: general overview

This first attempt to draw up a satellite account for enterprises in the social economy in Belgium consists of covering all the variables of production and generation of income accounts as well as the gross fixed capital formation on the capital account.

The different estimated aggregates for this exercise are therefore:

P1:   *Output*

P2:   *Intermediate consumption*

B1g: *Gross value added*

D1:   *Compensation of employees*

D29: *Other taxes on production*

D39: *Other subsidies on production*

B2g: *Gross operating surplus*

P51:   *Gross fixed capital formation*

The reference year is 2007 (in the general framework of the national accounts as published on 31 September 2009).

The estimation of aggregates of production and generation of income accounts generally proceeds in two main phases.

4.2.1. Administrative concepts

A first evaluation of aggregates on the basis of administrative and microeconomic data according to administrative concepts is realized. The calculation is made starting out from individual accounting data aggregated by branches of activity and by institutional sector.

Direct calculatory methods (based on accounts headings)[10] or *indirect* calculatory methods (referring to supplementary/alternative data sources[11] for the purpose of *ad hoc* extrapolation) were applied for each

---

[10]   Company annual accounts (submitted to the BNB Balance Sheet Centre) and/or exhaustive annual reports (in the case of financial companies).

[11]   Mainly National Social Security (NSS) salary data and data from Value Added Tax (VAT) declarations.

institutional sector and by branches of activity. Several categories of enterprises were thus mapped out according to the degree of precision and availability of accounting data. These are the enterprise categories A1, A2, B1, B2, B3, C1 and C2.

A1 → large enterprises with "complete" annual accounts

A2 → large enterprises without (usable) annual accounts

B1 → small and medium-sized enterprises (SMEs) with an abbreviated lay-out including the turnover and purchases, with a gross margin > 0

B2 → SMEs with an abbreviated lay-out without turnover or purchases, with a gross margin > 0

B3 → SMEs without (usable) annual accounts

C1 → SMEs with an abbreviated lay-out including the turnover and purchase, with a gross margin < 0

C2 → SMEs with an abbreviated lay-out without turnover or purchases, with gross margin < 0

The methods applied therefore differ according not only to institutional sector and branch of activity, but also category of enterprise.

4.2.2. Conversion to ESA 95 concepts

This essential step concerned making corrections to the administrative aggregates so as to bring them into line with the standards of the ESA 95.

For institutional sector S11[12] (non-financial corporations), the different types of correction recommended by the experts in the NAI are described in the *Inventaire* (pp. 80 to 99). They are based on information taken from annual accounts or, again, structure inquiry.

For institutional sector S12 (financial corporations) the necessary corrections vary according to the institutional subsector under review (S122, S123, S124 or S125). These have been processed by the experts at the BNB who sent us the variables adapted directly into ESA 95 concepts.

4.3. Aggregate calculation methods – Non-financial corporations (S11)

The calculations were made separately for each of our groups in population 1 (cooperatives) and 2 (CSPs) classified in institutional sector S11. To recall, Group 3 (mutual societies) falls exclusively within institutional

---

[12] Annex 3 presents the classification of institutional sectors.

sector S12. The method for calculating the aggregates for NACE branches agriculture and forestry departs from the general method.

4.3.1. Establishment of administrative aggregates (S11)

❖ *A direct calculatory method for enterprises in Category A1*

The large enterprises of Category A1 are those that submit their annual accounts to the Balance Sheet Centre according to the complete balance sheet lay-out. All the relevant variables for the calculation of "administrative" microeconomic aggregates are therefore available, namely:

Operating products

Code	Description
70	Turnover
71	Variation of stocks and goods produced
72	Internal production of fixed capital
74	Other operating products
740	Operating subsidies
741/9	Other miscellaneous operating products[13]

Operating costs

Code	Description
60	Supplies and merchandise
600/8	Purchase of merchandise, raw materials and goods[14]
609	Variation of stocks and goods
61	Purchases of miscellaneous goods and services (not entered in 600/8)
62	Remunerations, social contributions and pensions
64	Other operating costs
640	Fiscal operating costs
641/8	Other miscellaneous operating costs[15]

Setting out from this data we may deduce the following administrative microeconomic aggregates:

---

[13] 741/9 indicates the sum of the accounts from 741 to 749.
[14] 600/8 indicates the sum of the accounts from 600 to 608.
[15] 641/8 indicates the sum of the accounts from 641 to 648.

Administrative aggregate	Accounting code
(1) Output	70 + 71 + 72 + 74 - 740
(2) Intermediate consumption	60 + 61 + 641/8
(3) Personnel costs	62
(4) Net fiscal operating costs	640-740

And, on balance, we obtain the value added and the gross operating surplus (in "administrative" concepts):

Administrative balance	Accounting code
(5) Gross value added = [(1)-(2)]	70 + 71 + 72 + 74 - 740 - 60 - 61 - 641/8
(6) Gross operating surplus = [(5)-(3)-(4)]	70 + 71 + 72 + 74 - 60 - 61 - 62 - 640/8

❖ *An indirect calculatory method for enterprises in categories A2 to C2*

The methods used to estimate the administrative aggregates for categories A2, B1, B2, B3, C1 and C2 resort, in complement if not by default of (complete and usable) annual accounts data, to data taken from VAT declarations or from salary data from the National Social Security Office (NSSO).

4.3.2. Conversion to ESA 95 concepts (S11)

The next step in the construction of the satellite account consists of converting the microeconomic administrative aggregate into ESA 95 concepts.

To do so we must proceed by group of branches to corrections/reclassifications made in the relevant accountancy headings to recast them in ESA 95 concepts:

Microeconomic aggregates	Corrections / reclassifications	ESA 95 aggregates
70 + 71 + 72 + 74 - 740	============>	P. 1 Output
600/8 + 609 + 61 + 641/8	============>	P. 2 Intermediate consumption
62	============>	D.1 Compensation of employees
640	============>	D.29 Other taxes on production
740	============>	D.39 Other subsidies on production

The gross value added (B.1g) and the gross operating surplus (B.2g) in ESA 95 concepts are obtained from the balance.

In order to obtain ESA 95 aggregates, different corrections/reclassifications must be applied as recommended in the *Inventaire* (cf. Annex 4).

4.3.3. The particular case of enterprises of NACE branches agriculture and forestry

It should be emphasized that the general method described above does not apply to the branches NACE 01 (agriculture) and 02 (forestry). Estimation of the aggregates of these branches is in fact a matter of specific methods based on data relating to prices and to quantities. An *ad hoc* extrapolation method allowed estimation of the aggregates of our target population setting out from the aggregates of the whole of sector S11 for the branches concerned.

For agricultural activities the basic data on prices and produced quantities come from the *Directorate General Statistics and Economic Information (DGSEI),*[16] which also draws up the economic accounts for agriculture for Eurostat. Production (P. 1) must be estimated at basic price. Moving from the producer price to the basic price allows account to be taken of the large subsidies from which the sector benefits; it is necessary to add to production (valorized at producer price) the amount of subsidies on net products of taxes on products. Estimation of intermediate consumption (P. 2) is likewise determined by DGSEI.

Concerning forestry activities the data on the quantities produced come from the Walloon Region and are extrapolated for the whole of the national territory. The sale prices by species of wood and by quality of wood are available from the regions. In the absence of more precise information, the intermediate consumption is estimated in proportion with production.

4.4. Aggregate calculation methods – Financial corporations (S12)

The methods of evaluation of the aggregates of financial corporations differ according to the branch of activity to which they belong and according to institutional subsector under review.

The financial corporations of our population are for the most part spread between groups 3.1 (insurance mutuals) and 3.2 (health mutuals), both falling to be classified in institutional sector S125 and NACE branch 66.

---

[16] DGSEI is one of the directorates of the *Public Service Federal Economy, S.M.E., Middle Classes and Energy.*

4.4.1. Mutual insurances and community chests (Group 3.1)

The evaluation of aggregates of production and generation of income accounts for the *mutual insurance societies* and *insurance community chests under private law* (Group 3.1) was set out from:

- the structure of aggregates of companies in institutional subsector S125 and
- variable D.1 (compensation of employees) of our population.

Aggregates P. 1, P. 2, B.1g, B.2g, D.29 and D.39 were in fact extrapolated from the structure of the aggregates of companies in institutional subsector S125 (which is extracted from the national accounts for the year 2007).

4.4.2. Health mutuals (Group 3.2)

To recall, the health mutuals in Belgium are integrated in the national social security system and are closely associated with the management of compulsory health insurance. This field of their activities does not answer to the principle of autonomy of management so dear to the social economy and should be excluded from it. However, they develop – in parallel and autonomously – complementary insurances and various social services that integrate perfectly in our field of analysis.

So, with a view to evaluation of aggregates of production and operation accounts of our population of health mutuals, we now have to isolate the sole activities linked to optional complementary insurance or other social services (excluding any activity of a compulsory nature).

The evaluation of aggregates is based on exhaustive data taken from the 2008 Annual Report of the *Office de contrôle des mutualités et des unions nationales des mutualités*[17] (OCM). The financial statements for the 2007 financial year are used.

More precisely, the basic data is taken from the *Compte de résultats général de l'assurance libre et complémentaire et de l'épargne prénuptiale – situation globalisée des entités mutualistes –* (cf. OCM Report, pp. 114). From this globalized account we then deduce:

- (i), the data for *Health Care Insurance*, compulsory in the Flemish Region,[18] this insurance is designed to defray the costs of non-medical aid and services for persons affected by prolonged reduction of their autonomy (cf. OCM Report, p. 99); and

---

[17] Office de contrôle des mutualités et des unions nationales des mutualités, *Annual Report,* Brussels, 2008, p. 162.

[18] Insurance introduced by the Flemish Community Decree of 30 March 1999.

- (ii), the data for the *Fonds spécial de réserve complémentaire*,[19] legal reserve introduced in the framework of compulsory insurance.

4.5. Calculation of Gross fixed capital formation (P51)

Estimation of aggregate P51, in accordance with the conceptual framework of national accounts, is in the main based on data taken from annual accounts (Balance Sheet Centre) or, failing that, from VAT declarations. Data from the structural inquiry is also used sometimes.

For our population it was possible to use an *ad hoc* method based on an adjustment coefficient for the administrative aggregates proper to each group of activity; and, by default (where the data is unsuitable for this), we used an extrapolation ratio based on compensation of employees (D1).

5. Results

This section presents the results of the first drawing up of the satellite account of enterprises in the social economy in Belgium. For the year 2007 this covers the sequence of aggregates of production and generation of income accounts (Section 5.1) and the aggregate P51 (gross fixed capital formation) of the capital account (Section 5.2).

5.1. Aggregates of production and generation of income account by population group, by institutional sector and by branch

5.1.1. Results per group of population

As shown in Table 5/1, the enterprises of our population account for a total production of 4,149 million euros. Nearly 59% of that production (2,435 million euros) comes from the group of accredited cooperatives, 20% from the mutual insurance societies and community chests (IM & CC), 16% from the CSPs and, finally, 5% from the mutual health societies (non-compulsory activities).

However, although the group of accredited cooperatives is largely in the lead in terms of production, it cedes its first place to the IM & CC group in terms of value added and, again, to the CSPs in terms of gross operating surplus (under the crushing weight of operating costs, as we shall emphasize later).

---

[19] Art. 199 of the coordinated law relating to compulsory insurance of 14 July 1994, introducing the obligation to charge a fee with a view to building up the reserve funds in 1995 and 1996. Starting in the 1997 financial year the insurance organizations have continued to charge this fee as a prudential measure.

So, with a gross value added exceeding 326 million euros out of a total of 942 million euros for the whole of our population, the IM & CC group alone realized nearly 35% of the total value added. The 'NCC' cooperatives contribute at the rate of 26%, the CSPs 25% and, finally, the health mutuals 14%.

The operating costs (not including fixed capital consumption) of our productive units totaled 3,661 million euros, of which 3,207 million euros (88%) in intermediate consumptions of goods and services, 617 million euros (17%) in compensation of employees and -163 million euros (-4%) in other net taxes on subsidies.

This concise analysis of the composition of operating costs thus reveals that it is the intermediate purchases of goods and services that weigh hardest on the enterprises of our population. This is quite particularly true for the 'NCC' cooperatives of our population, alone carrying 68% of the total amount of intermediate consumptions in the population (for a value of 2,187 million euros), the mutuals and CSPs carrying respectively 16% and 14% of the item in question.

Salaries and wages for all the groups taken together represent 617 million euros, respectively divided 42%, 22% and 37% between the accredited cooperatives, the CSPs and the mutuals.

The other net taxes on subsidies are globally negative. This situation results from aid received by the sector globally, and more particularly from subsidies entered into accounts under the assets of health mutuals (non-compulsory activities) for which a zero amount is attributed to 'other taxes on production'.

5.1.2. Results per institutional sector

It should first be pointed out that institutional sector S11 (non-financial corporations) covers almost all our 'NCC' cooperatives and CSPs (groups 1 and 2), whereas the financial corporations sector (S12) is made up almost exclusively of our mutuals classified within groups 3.1 and 3.2; these latter are superposed on each other in subsector S125 (insurance societies and pension funds). The results presented per institutional sector therefore remain very similar to the findings shown previously in the per population group analysis.

Presentation per institutional sector nevertheless sheds more light on size: it allows replacement of the satellite account of enterprises of the social economy in the framework of national accounts by direct comparison of the results for our population against those for corresponding institutional sectors, and this at national level.

Table 5/2 present those results. We may say that, all in all, the companies of our target population contribute 0.7% to the production of all the companies taken as a whole nationwide (S11 and S12) and make a 0.5% contribution to the gross value added of all the companies of the country (S11 and S12).

The contribution from enterprises of the social economy that falls to be classified in institutional sector S125 is especially striking. In fact, these enterprises alone realize more than 12% of the production of sector S125 at national level and generate 14.5% of the value added of sector S125 of the nation.

For more details, the distribution of our population groups within each institutional sector (S11 and S12) is presented in Annex 5.

5.1.3. Results per branch

The majority of enterprises of the social economy are present in the tertiary sector (Code 3 of NACE classification A3). This is good for up to 92% of the total value added produced by our population; 78% of total production; 91% of compensations of employees.

In the tertiary sector, i.e., Branch 5 (NACE Classification A6) grouping financial and real estate activities and services to businesses are to the fore, followed by Branch 4 (commerce, repair of motor vehicles, hotels and restaurants, transport and communication).

It should be noted that the secondary sector (Code 2, Classification A3) and primary sector (Code 1, A3) represent respectively nearly 7% and 0.6% of the total wealth created by our population in terms of value added.

Among these secondary-sector activities, the manufacturing industry and energy branch (Code 2 of Classification A6) contributes 6.5% to the total value added of our population, whereas the construction sector (Code 3, A6) contributes 0.7%.

Table 5/1. *Production and generation of income accounts by population group and in total, estimations at current prices, 2007 (thousands euros)*

	P1	P2	B1g	D1	D29	D39	B2g
Group 1: 'NCC' Cooperatives	2,435,428	2,187,043	248,385	258,053	4,811	62,454	47,975
Group 2: CSPs	678,027	442,466	235,561	133,581	6,111	57,382	153,252
Subtotal groups 1 & 2	*3,113,455*	*2,629,509*	*483,946*	*391,634*	*10,922*	*119,836*	*201,227*
Group 3.1 IM and CC	841,208	514,761	326,447	188,666	5,448	0	132,333

	P1	P2	B1g	D1	D29	D39	B2g
Group 3.2 Health mutuals (non-compulsory activities)	194,283	63,171	131,112	37,049	0	60,260	154,323
Subtotal Mutuals (groups 3.1 & 3.2)	*1,035,491*	*577,932*	*457,559*	*225,715*	*5,448*	*60,260*	*286,656*
Overall total	**4,148,946**	**3,207,441**	**941,505**	**617,349**	**16,369**	**180,096**	**487,883**

Note: As a reminder, the codes of the ESA 1995 headings mean:
P1: Output
P2: Intermediate consumption
B1g: Gross value added
D1: Compensation of employees
D29: Other taxes on production
D39: Other subsidies on production
B2g: Gross operating surplus

Table 5/2. *Production and generation of income accounts by institutional sector and in total, estimations at current prices, 2007 (thousands euros, %)*

	P1	P2	B1g	D1	D29	D39	B2g
S11	**3,027,382**	**2,546,076**	**481,307**	**360,906**	**10,484**	**119,805**	**229,722**
in % S11 nation	0.5%	0.7%	0.3%	0.3%	0.5%	2.9%	0.3%
S12	**1,121,564**	**661,365**	**460,199**	**256,443**	**5,886**	**60,292**	**258,162**
among which S125	1,035,491	577,932	457,559	225,715	5,448	60,260	286,656
in % total S125 nation	12.5%	11.4%	14.2%	12.1%	10.1%	99.9%	21.0%
in % total S12 nation	3.2%	3.5%	2.9%	2.8%	1.5%	82.4%	4.0%
Overall total	**4,148,946**	**3,207,441**	**941,505**	**617,349**	**16,369**	**180,096**	**487,883**
in % total nation (S11+S12)	*0.7%*	*0.8%*	*0.5%*	*0.5%*	*0.6%*	*4.3%*	*0.6%*

Note: As reminder, according to ESA 1995 classification of the institutional sectors, following codes mean:
– S11: non-financial corporations,
– S12: financial corporations,
– S125: insurance corporations and pension funds.

Table 5/3. *Production and generation of income accounts by activity branch (A3 and A6), estimations at current prices, 2007 (thousands euros)*

A3	A6	P1	P2	B1g	D1	D29	D39	B2g
1		**15,684**	**9,755**	**5,930**	**4,715**	**25**	**1,122**	**2,312**
	1	15,684	9,755	5,930	4,715	25	1,122	2,312
2		**760,990**	**693,054**	**67,936**	**46,913**	**939**	**614**	**20,697**

	2	736,027	674,931	61,096	43,762	790	449	16,993
	3	24,963	18,123	6,840	3,151	149	165	3,705
3		**3,372,272**	**2,504,632**	**867,640**	**565,721**	**15,406**	**178,3606**	**464,873**
	4	683,935	500,478	183,457	145,365	2,021	7,676	43,747
	5	2,387,829	1,750,954	636,876	362,946	13,033	135,3428	396,239
	6	300,507	253,200	47,307	57,410	352	35,342	24,887
Overall total		**4,148,946**	**3,207,441**	**941,505**	**617,349**	**16,369**	**180,096**	**487,883**

Note: the description of NACE A3 and A6 is to be found in Annex 2.

5.2. Aggregate P51 – Results by population group, by institutional sector and by branch

Tables 5/4 and 5/5 show the distribution of the Gross Fixed Capital Formation (GFCF) respectively by population group and by institutional sector.

Table 5/4. *Distribution of gross fixed capital formation by population group and in total, estimations at current prices, 2007 (thousands euros, %)*

	P51	%
Group 1: 'NCC' Cooperatives	82,989	35
Group 2: SFS	97,842	42
Subtotal groups 1 & 2	*180,831*	*77*
Group 3.1: IM and CC	45,663	19
Group 3.2: Health mutuals (non mandatory activities)	8,166	3
Subtotal Mutuals (groups 3.1 & 3.2)	*53,829*	*23*
Overall total	**234,661**	**100**

Table 5/5. *Distribution of gross fixed capital formation by institutional sector and in total, estimations at current prices, 2007 (thousands euros, %)*

	P51
S. 11	**178,759**
in % S11 nation	*0.4%*
S. 12	**55,902**
among which S. 125	*53,829*
in % total S125 nation	**15.4%**
in % total S12 nation	*2.1%*
Overall total	**234,661**
in % total nation (S11+S12)	*0.5%*

Sources: ICN 2010, own calculations.

193

The GFCF of our productive units taken as a whole is estimated at nearly 235 million euros. As Table 5/4 shows, 42% of that sum comes from the CSPs, 35% from the cooperatives and 22% from the mutuals.

The distribution by institutional sector, as shown in Table 5/5, reveals among other things the predominance of investments from the non-financial corporations (S11) compared with the financial corporations (S12): 73% of the total GFCF is realized by S11 as against 24% by S12.

This presentation of the GFCF by institutional sector also allows replacement of the indicator in question in the conceptual framework of national accounts. We may also note that the companies of the social economy as a whole contribute 0.5% to the investments by companies nationwide. More particularly again, we call attention to the considerable contribution from companies in sector S125: these contribute 15% to the formation of fixed capital for the whole subsector at national level.

The results given above are perfectly consistent with the production indicators and, more particularly, with aggregates P1 and B1g as, incidentally, was only to be expected.

The detail of the distribution of investments of our productive units per branch of activity is presented in Table 5/6 at different levels of aggregation (NACE A3 and A6).

Table 5/6. *Distribution of gross fixed capital formation by activity branch (A3 and A6), estimations at current prices, 2007 (thousands euros, %)*

A3	A6	P51	%
1		31,329	13.4
	1	31,329	13.4
2		10,008	4.3
	2	3,498	1.5
	3	6,511	2.8
3		193,324	82.4
	4	24,444	10.4
	5	151,774	64.7
	6	17,105	7.3
Overall total		234,661	100

Note: the description of NACE A3 and A6 codes can be found in Annex 2.

First of all we confirm the predominance of the tertiary sector, with 82% of the total of investments by the whole of the enterprises in our population; this again matches the previous findings for aggregate P1, production; (to recall, 78% of the total production came from the tertiary sector). However, there is a marked division of the remainder between primary and secondary sector: in fact, whereas the primary sector contributes 0.6% of the total in terms of production, it bags more than 13% of the total in terms of GFCF. This result is not unrelated to the actual structure of production of agricultural enterprises (and, more particularly, with the presence of farmer cooperatives akin to the cooperatives for the common use of agricultural equipment, *coopératives d'utilisation en commun de matériel agricole* – CUMA).

Among the secondary activities here it is the activities of the Construction branch (Code 3 Classification A6) that predominate; this is readily explained by the very nature of the activity.

Finally, among the tertiary activities, the financial, real estate, renting and business activities (Code 5 Classification A6) form the majority, representing 64.7% of investments.

6. Conclusion

The SATACBEL project presented in this chapter has allowed a first experience in Belgium with the satellite account.

6.1. A juridico-institutional referencing

We used a juridico-institutional referencing to delimit our population.

The concept of the social economy has been the subject of many a debate between experts and scientists throughout the world. The statistical contours of the social economy will thus depend on preferred conception. In its widest and its most generally accepted meaning, *the social economy consists of economic activities producing goods or services, conducted by societies, mainly cooperatives and/or companies with social purpose, associations, mutual societies or foundations, whose ethic finds expression in four principles (objective of service in the group or general interest, or for the members, rather than mere pursuit of profit; autonomy of management; democratic decision-making process; primacy of people and labour over capital in the distribution of income).*

It is this definition that has been chosen as the basis for delimitation of the field of the present study and that guided our choice for inclusion or exclusion whilst drawing a demarcation line between the public sector and the for-profit commercial private sector.

In Belgium, four main types of organizations devote themselves to the principles of definition of the social economy. These are the:

- associations (non-profit institutions, foundations, unincorporated associations),
- cooperative societies pursuing a genuine cooperative plan,

- societies adopting the form of a company with social purpose, and
- mutual societies.

Since our satellite account has the objective of evaluating the economic weight of the mutuals and cooperatives and/or societies with a social purpose whilst avoiding any useless reduplication with the satellite account of the NPIs – already the subject of regular calculations – the associations are excluded from our population.

This leads us to take as target population the three other organizational components, provided, of course, that they remain faithful to the four basic principles of the social economy. In Belgium different status and laws have been introduced over time. Thus seeing the emergence of cooperatives having chosen this status for no reason other than legal convenience, the legislator reacted by creating the National Cooperation Council (NCC), the accreditation from the NCC attesting that the principles of the social economy are truly respected. More recently the label of social purpose (that may be adopted by different commercial companies already in existence) was born of the will to bridge a certain legal gap, allowing a combination of social purpose and commercial activity.

Taking into account those juridico-institutional specificities, the next step has consisted in drawing up lists of companies making up our target population based on our juridico-institutional system. After cleaning those individual data (repeats, reclassification, …), our study ultimately covered a population of: 461 cooperatives accredited by the National Cooperation Council (*Group 1: 'NCC' coop*), 421 companies with social purpose (*Group 2: 'CSPs'*), 18 mutual insurance societies and community chests (*Group 3.1: IM & CC*) and 5 national unions of health mutuals (including all the mutualist entities that they cover, but excluding their compulsory-insurance activities) (*Group 3.2*).

6.2. Methodological contribution

This work must be conceived, if anything, as an important methodological advance since, up to now, no homogeneous method had ever been implemented at Belgian national and European level to analyze this particular field of the social economy.

To guarantee coherence with the central framework of national accounts and with the methodology used for the drawing up of satellite accounts for the NPIs in Belgium, it was important to be very close to the methodology of the National Accounts Institute (NAI) within the Belgian National Bank (BNB). For this reason, the project was conducted in close cooperation with the experts of the NAI and the BNB.

Bearing in mind the timeline for the project and the complexity of construction of accounts, it became apparent that – for macroeconomic analysis purposes – only the more significant indicators could be calculated (production, intermediate consumption, value added, compensation of employees, gross operating surplus and gross fixed capital formation).

The SATACBEL project also offers a significant statistical aid, giving the first sets of figures in Belgium for the key macroeconomic variables by type of enterprise (cooperatives, mutuals and CSPs), by institutional sector and by branch. However, the data thus obtained must be used with the utmost caution in terms of international comparison, since the demography of the studied populations varies greatly from country to country. Despite the methodological and statistical limitations, this satellite account allows greater visibility and recognition of the field of the social economy and constitutes a valuable tool for use by political decision-makers, the managers in the field and the specialists who study the field. Its replication in time and its extension to other aggregates should allow not only the systematic count of these enterprises and the evaluation – through various indicators – of their real economic weight, but also the cartography of the branches of activity in which they operate.

The continued development and extension of this statistical tool should contribute significantly to the sustained efforts of the organizations of the social economy, not only Belgian but also European, coupled with those of the Belgian public authorities and European institutions, to give greater institutional visibility to the sector of the social economy. National and European policies would thus benefit from new light shed on a series of questions. Mention may be made here, for instance, of the evaluation of the necessity of a European status for mutual societies, of the model of the European cooperatives status; the best taking into account of the cooperative and mutualist models as forms of doing business, having to answer to the same conditions as any other enterprise of the rules of competition; the recognition of the active role of the cooperatives and the mutual societies in the European policy of social cohesion.

References

AISAM, *L'assurance mutuelle: qu'est ce que c'est? Pourquoi faire? Un guide pour les sociétaires et les collaborateurs*, Brussels, Edition Lieve Lowet, AISAM, http://www.amice-eu.org/userfiles/file/AISAM_What_is_Mutuality_fr.pdf.

Archambault, E., *Comptabilité nationale*, 6e edition, Paris, Economica, 2003.

Barea, J., Monzón, J. L., *Manual for Drawing up the Satellite Accounts of Companies in the Social Economy: Co-operatives and Mutual Societies*, Liège, CIRIEC, 2006.

Ben Sedrine, W., Fecher, F., Sak, B., *Comptes satellites pour les coopératives et mutuelles en Belgique. Première élaboration* (SATACBEL), Liège, CIRIEC, 2011.

D'Hulstère, D., Pollénus, J.-P., *La société à finalité sociale en questions et réponses*, Liège, Edipro, 2008.

European Commission, *Le Secteur coopératif, mutualiste et associatif dans l'Union européenne*, Luxembourg, Office des publications officielles des Communautés européennes, 1997.

Eurostat, *Système européen des comptes nationaux et régionaux: SEC 1995*, Eurostat, 1996, http://circa.europa.eu/irc/dsis/nfaccount/info/data/esa95/fr/titelfr.htm.

Institut des Comptes Nationaux, *Comptes nationaux. Le compte satellite des institutions sans but lucratif 2000-2004*, Bruxelles, Banque Nationale de Belgique, juin 2007.

Institut des Comptes Nationaux, *Comptes nationaux. Partie 2 – Comptes détaillés et tableaux 1999-2008*, Bruxelles, Banque Nationale de Belgique, octobre 2009.

Institut des Comptes Nationaux, *Méthode de calcul du produit intérieur brut et du revenu national brut selon le SEC 1995*, http://www.nbb.be/doc/DQ/F_method/M_Inventaire_SEC1995_FR_def.pdf.

United Nations, *Handbook on Nonprofit Institutions in the System of National Accounts*, New York, United Nations, 2003.

Van Opstal, W., Gijselinckx, C., Develtere, P. (eds.), *Entrepreneuriat coopératif en Belgique, Théories et pratiques*, Leuven, Acco, 2008.

Annexes

Annex 1. The nine principles of the companies that pursue a social purpose (CSPs)

"Companies with the legal personalities mentioned in Article 2, § 2, exception being made in respect of the Societas Europea (European Company) and of the European Cooperative Society, are referred to as companies with social purpose if they are not driven by the enrichment of their members and where their Articles of Association:

1° stipulate that members shall seek only a limited patrimonial profit or no patrimonial profit at all;

2° define precisely the ends to which the activities referred to in their social purpose shall be dedicated and not take as principal objective of the company the creation of indirect profits on assets for members;

3° define the policy of allocation of profits in accordance with the internal and external objectives of the company, in accordance with the hierarchy established in the Articles of Association of the said society, and the policy of building up of reserves;

4° stipulate that no-one may take part in the vote on the general assembly for a number of votes exceeding one tenth of the votes attached to represented shares; this percentage is reduced to one twentieth when one or more members are acting in the capacity of member of personnel engaged by the company;

5° stipulate, where the society creates a direct limited profit for its members, that the profit distributed to them may not exceed the interest rate set by the His Majesty the King in execution of the Law of 20 July 1955, establishing a National Cooperation Council applied to the amount actually released in 'parts sociales' and shares;

6° provide that, each year, the administrators or managers shall submit a special report on the way in which the society sought to achieve the objective that it has set for itself in accordance with point 2; this report shall establish, inter alia, that the expenditure concerning investments, operating costs and remunerations is conceived in such a way as to favour the achievement of the objective of the company;

7° make provision for special rules and procedures allowing each member of the personnel to acquire, not later than one year after engagement by the company, the status of associate; this provision does not apply to members of the personnel who do not enjoy full civil capacity;

8° make provision for special rules and procedures whereby members of personnel no longer bound to the company by an employment contract shall lose the status of associate not later than one year of termination of such contractual bond;

9° stipulate that, after settlement of any liabilities and the reimbursement of their presentation to the members, the surplus from liquidation shall receive an allocation that most nearly approximates the social purpose of the company."

Annex 2. Classifications of activity branches (NACE 2003)

Classification A3 – (NACE reference)

Code	Description of branch	NACE reference Rev. 1
1	AGRICULTURE, HUNTING, FORESTRY and LOGGING; FISHING; and AQUACULTURE	A + B
2	INDUSTRY, INCLUDING ENERGY and CONSTRUCTION	C + D + E + F
3	SERVICES ACTIVITIES	G to P

Classification A6 – (NACE reference)

Code	Description of branch	NACE reference Rev. 1
1	AGRICULTURE, HUNTING, FORESTRY and LOGGING; FISHING; and AQUACULTURE	A + B
2	INDUSTRY, INCLUDING ENERGY	C + D + E
3	CONSTRUCTION	F
4	WHOLESALE AND RETAIL TRADE; REPAIR OF MOTOR VEHICLES, MOTORCYCLES AND PERSONAL AND HOUSEHOLD GOODS; HOTELS AND RESTAURANTS; TRANSPORT, STORAGE AND COMMUNICATIONS	G + H + I
5	FINANCIAL, REAL ESTATE, RENTING AND BUSINESS ACTIVITIES	J + K
6	OTHER SERVICES ACTIVITIES	L to P

Classification A60 – (NACE reference)

Code_Value	Code_Description
C01	Products of agriculture, hunting and related services
C02	Products of forestry, logging and related services
C05	Fish and other fishing products, services incidental to fishing
C10	Coal and lignite; peat
C11	Crude petroleum and natural gas; services incidental to oil and gas extraction excluding surveying
C12	Uranium and thorium ores
C13	Metal ores
C14	Other mining and quarrying products
C15	Food products and beverages
C16	Tobacco products
C17	Textiles
C18	Wearing apparel; furs
C19	Leather and leather products
C20	Wood and products of wood and cork (except furniture), articles of straw and plaiting materials
C21	Pulp, paper and paper products
C22	Printed matter and recorded media
C23	Coke, refined petroleum products and nuclear fuel
C24	Chemicals; chemical products and man-made fibres
C25	Rubber and plastic products
C26	Other non-metallic mineral products
C27	Basic metals
C28	Fabricated metal products, except machinery and equipment
C29	Machinery and equipment n.e.c.
C30	Office machinery and computers

C31	Electrical machinery and apparatus n.e.c.
C32	Radio, television and communication equipment and apparatus
C33	Medical, precision and optical instruments, watches and clocks
C34	Motor vehicles, trailers and semi-trailers
C35	Other transport equipment
C36	Furniture, other manufactured goods n.e.c.
C37	Recovered secondary raw materials
C40	Electrical energy, gas, steam and hot water
C41	Collected and purified water, distribution services of water
C45	Construction work
C50	Trade, maintenance and repair services of motor vehicles and motorcycles; retail trade services of automotive fuel
C51	Wholesale trade and commision trade services, except of motor vehicles and motorcycles
C52	Retail trade services, except of motor vehicles and motorcycles; repair services of personal and household goods
C55	Hotel and restaurant services
C60	Land transport and transport via pipeline services
C61	Water transport services
C62	Air transport services
C63	Supporting and auxiliary transport services; travel agency services
C64	Post and telecommunication services
C65	Financial intermediation services, except insurance and pension funding services
C66	Insurance and pension funding services, except compulsory social security services
C67	Services auxiliary to financial intermediation
C70	Real estate services
C71	Renting services of machinery and equipment without operator and of personal and household goods
C72	Computer and related services
C73	Research and development services
C74	Other business services
C75	Public administration and defence services; compulsory social security services
C80	Education services
C85	Health and social work services
C90	Sewage and refuse disposal services, sanitation and similar services
C91	Membership organization services n.e.c.
C92	Recreational, cultural and sporting services
C93	Other services
C95	Private households with employed persons
C99	Services provided by extra-territorial organizations and bodies

Annex 3. Classification of institutional sectors (ESA 1995)

S. 1 Total Economy

S. 11 Non-Financial Corporations

S. 12 Financial Corporations

S. 121 The Central Bank

S. 122 Other monetary financial institutions

S. 123 Other financial intermediairies, except insurance corporations and pension funds

S. 124 Financial auxiliaries

S. 125 Insurance corporations and pension funds

S. 13 General Government

S. 14 Households

S. 15 Non-Profit Institutions Serving Households

S. 2 Rest of the world

Annex 4. Corrections for conversion to ESA 1995 concepts

	Type of correction	Heading on the basis of which the correction coefficient is calculated
(a)	Removal of taxes on products and non-deductible VAT from "fiscal operating costs"	640
(b)	Alignment of salaries of annual accounts with the salary mass of ESA 95	62
(c)	Transfer of rebates for cash payment of profit or loss on turnover and purchases	70
(d)	Removal of purchased products from turnover and purchases	70
(e)	Removal or current capital gains or losses from other products and operating costs	74-740
(f)	Removal of rents collected and paid from areas turnover and purchases	74-740
(g)	Removal of R&D developed internally from P. 1 and transfer of investments in purchased R&D to P. 2	72
(h)	Removal of donations from purchases	600/8+61
(i)	Activation of software packages purchased or developed internally	72
(j)	Transfer of formation expenses entered under assets to intermediate consumption	600/8+61
(k)	Transfer of certain bank charges from financial costs to purchases of services	600/8+61
(l)	Removal of the part of transfer of damage insurance premiums paid on purchases	600/8+61

(m)	Removal of damage insurance compensation received from other operating products	74-740
(n)	Transfer of subsidies of interests received from financial products to operating subsidies	740
(o1)	Removal of excises on purchases, fiscal operating costs and turnover	70
(o2)	Removal of turnover tax, fiscal operating costs and turnover in the pharmaceuticals industry	N/A
(o3)	Removal of other taxes on production entered under turnover	740
(o4)	Transfer to turnover of taxes on products entered under fiscal operating costs	740
(p1)	Salaries in natura produced internally (increase of turnover)	62
(p2)	Purchased salaries in natura (removal of purchases)	62
(q)	Recording of gratuities paid in turnover and salaries	70
(r)	Transfer of shares paid out from profits to be distributed in purchases of services	600/8 61
(s)	Removal of purchased immovable assets intended for sale from turnover and purchases	70
(t)	Transfer of variations of stocks and investment assets produced internally to turnover	70
(v)	Removal of value added produced in a foreign country	N/A
(w)	Removal of stock evaluation differences	C_C (value added)
(x)	Additions (hospitals, original works, housing services, own accommodation, home services staff)	N/A
(y)	Increased charge for undeclared employment	C_C (value added)
(z)	Corrections in purchases and turnover beyond goods (public sales, games of chance, energy)	C_C
(aa)	Removal of difference between D.29/D.39 paid and received and insurance premiums	640
(ab)	Redistribution of protected workshops	N/A
(ac)	Grossing-up of outsourcing	N/A
(ad)	Intermediation between perspective of production and perspective of expenditure	640
SIFIM	Reallocation of intermediate consumption from FISIM to FISIM activity branches	C_B (intermediate consumption)

Notes: - N/A: correction not adopted for the study population;
- FISIM: Financial Intermediation Services Indirectly Measured;
- C_C: gross value added in administrative concept (as described in Table 4/1);
- C_B: intermediate consumption in administrative concept (as described in Table 4/1).

Annex 5. Production and generation of income accounts by institutional sector and in total – Distribution by group of population, estimations at current prices, 2007 (thousands euros, %)

	P1	P2	B1g	D1	D29	D39	B2g
S11	**3,027,382**	**2,546,076**	**481,307**	**360,906**	**10,484**	**119,805**	**229,722**
Group 1 'NCC' cooperatives	2,349,355	2,103,610	245,745	227,325	4,373	62,422	76,470
% population S11	78%	83%	51%	63%	42%	52%	33%
Group 2 CSPs	678,027	442,466	235,561	133,581	6,111	57,382	153,252
% population S11	22%	17%	49%	37%	58%	48%	67%
Total S11 (%)	100%	100%	100%	100%	100%	100%	100%
S12	**1,121,564**	**661,365**	**460,199**	**256,443**	**5,886**	**60,292**	**258,162**
Group 3.1 mutual insurances	841,208	514,761	326,447	188,666	5,448	0	132,333
% population S12	75%	78%	71%	74%	93%	0%	51%
Group 3.2 health mutuals	194,283	63,171	131,112	37,049	0	60,260	154,323
% population S12	17%	10%	28%	14%	0%	100%	60%
Group 1 'NCC' cooperatives	86,073	83,433	2,640	30,728	438	32	-28,495
% population S12	8%	13%	1%	12%	7%	0%	-11%
Total S12 (%)	100%	100%	100%	100%	100%	100%	100%
Overall total	**4,148,946**	**3,207,441**	**941,505**	**617,349**	**16,369**	**180,096**	**487,883**

Notes: Only 2 CSPs belong to sector S12; they were globalised with 'NCC' cooperatives of group 1.

Collaborative Research Between Civil Society, State and the Academia

Lessons from the Brazilian Mapping of the Solidarity Economy

Full Professor at the Universidade do Vale do Rio dos Sinos, Brazil

Introduction

The social and solidarity economy is increasingly receiving interest in both the North and the South. Activists and researchers value its component of reciprocity and its expediency to spread democracy beyond countries and continents through autonomous public spaces created by civil society. As such, the social and solidarity economy has become a major topic in the debates on current issues of society on a national and worldwide scale.

In the North, the social and solidarity economy has its roots primarily in France, Belgium and Spain, from where it spread in Quebec, Canada. It comprises a set of collective initiatives seeking to establish autonomous and democratic forms of management. In these initiatives, the ways in which power is shared and income distributed result from the primacy of people over capital and from the aim pursued – namely of providing a service to the members and to the community to which they belong. Economic activity and its surplus are, therefore, a means rather than an end in and of itself, or a way of making profit.

In Latin America, the solidarity economy usually refers to economic organizations aiming not only for financial gain for their members but also for benefits in terms of quality of life and citizen participation. Such goals are achieved mainly through significant efforts that include the engagement of the associated workers. Because of their social and community embeddedness, these initiatives also fulfill a number of functions in the fields of health, education and environmental protection, among other areas. Their associative economic practices are characterized by their commitment to democratic principles and cooperation and involve pooling the means of production, the work processes and resources from management.

The solidarity economy in Latin America encompasses a multitude of social segments, agents and institutions. As a global movement, it engages in theoretical criticism of the capitalist economic system; however, overall it is more practice-oriented and focuses its efforts on supporting enterprises in achieving individual, social and ecological goals. These enterprises include cooperative banks, service and goods exchange based on reciprocity, commercial networks and, above all, countless informal or formal associations of people who freely come together with the goal of developing economic activities, creating jobs and experimenting solidarity-based relations, be it among themselves or in society at large.

In Brazil, non-profit and self-managed experiences are an indelible mark of the solidarity economy. The boom in the number of initiatives observed over the last years can partly be attributed to the structural crisis that hit the Brazilian labor market and whose impacts were reinforced by the withdrawal of the state. Another important factor accounting for the boom of the solidarity economy is the mobilization of social movements, labor unions and citizen entities, unwavering in their commitment to establish and foster mutual help and economic cooperation practices. Both in the countryside and in suburbs, thousands of small community-based initiatives have been taking root for a long time, introducing the solidarity economic practices that have since spread and gained broader recognition.

The solidarity economy in Latin America, as well as the social economy in the North, corresponds to phenomena linked to specific contexts and periods. In Brazil, for example, a country with a fairly deep-rooted, although not always continuous, history of solidarity-oriented values, it has been emerging gradually since the 1980s. This history includes many manifestations of solidarity with workers, even if these could differ significantly depending on the national and regional context. Thus, aware of those differences, we are less eager about finding a universal, one-size-fits all concept and more interested in exploring how the solidarity economy could be best implemented in specific contexts.

In Brazil, this remains an important and challenging task. In April 2013, the second National Mapping of the Solidarity Economy was completed. This undertaking involved visits throughout the country to identify *solidarity economy enterprises*. Nearly 20,000 enterprises were identified, the majority of which were associations (59.9%), followed by informal groups (30%) and, to a much smaller extent, cooperatives (8.8%). Although this second Mapping was much delayed (its completion was scheduled for 2010) and although it fell short of initial targets of mapping over 30,000 enterprises, it constituted a collaborative research of paramount importance. Overall, both the first (2005-2007) and the second (2009-2013) National Mapping are symptomatic of the gaps in the current state of knowledge on the solidarity economy in Brazil. Nevertheless, given the body of empirical data they have collected and the unique methodology they developed, they are valuable for overcoming such deficiencies.

The first main knowledge gap involves the lack of statistical information about the typical organizations of the solidarity economy in Brazil. Available surveys, apart from the Mappings, do not provide a broad base of information and were carried out without continuity and systemization, preventing a cross-comparison of data. Thus, except for the Mappings, very little can be said about the statistical population of the solidarity economy or the characterization of the main forms of solidarity enterprises. The inconsistency of statistics hinders even the analysis of those sectors that have a stable regulatory framework and a reasonable degree of institutionalization, such as cooperatives.

Secondly, even if there were systematic and comprehensive statistics on the most common forms of organization of the solidarity economy, under the prevailing conditions of the available surveys, such statistics would be inappropriate in most cases. This is because solidarity economicy enterprises tend to adopt one of the available institutional structures – usually an *association* or *cooperative* – only because they lack alternatives more suited to their goals and their *sui generis* dynamics. Thus, they view such structures as stopgap solutions to avoid informality. Therefore, the solidarity economy does not identify itself with the associative sector or cooperatives, although it borrows their organizational modalities. Instead, it supports the unanimous request that the Brazilian regulatory framework comprise new juridical forms that better reflect the concrete experiences and concepts generated from this field of practices.

Thirdly, since there is no consensus on the most suitable legal frameworks, any comprehensive data survey on the solidarity economy must develop its own criteria for defining its target population. Indeed, one of the greatest challenges of the Mapping was to establish, for the first time, a set of classification parameters of economic organizations and similar *solidarity-based* initiatives. This challenge, methodological and political at the same time, gave rise to a series of fierce debates before and after the Mapping. And, although important steps were taken, unanimous decisions were not reached. Social actors continue to have contradictory perceptions and intervention processes, with the effect that certain organizations are included in the solidarity economy in some instances yet excluded in other instances. Subsequent to the first Mapping, the prevailing perspective was to enlarge the frontiers of the solidarity economy. The second Mapping reinforced this trend, with the advantage of providing a more accurate understanding of the internal diversity of the solidarity economy and its interconnections with similar and surrounding organizations.

This inclusive recognition process assumes that the field of the solidarity economy should be carefully analyzed. Which leads to the fourth challenge raised by the Mappings, namely that there has hardly been, thus far, a debate about solidarity economy indicators and statistics in Brazil. One sign of that is the scarce analytical interest raised by the first Mapping, even though many universities contributed to its production. The underlying problem here is, in turn, the predominance of qualitative studies that are limited to a casuistic approach of such alternative economic experiences. More specifically, they focus on cases confined to their particular circumstance and whose analysis gives a decisive value to the direct relationship between the researcher and the realities under study. When conducted with a rigorous methodology, such analyses can indeed capture relevant singularities but nevertheless fail to allow for a generalization of results or to contribute significantly to the evaluation of the predominant traits of the solidarity economy, including an identification of its trends or an assessment of the most important obstacles and propelling factors.

These comprise the main causes of the inconclusive state of knowledge on the solidarity economy in Brazil – a state that has either paralyzed debates or made for counterproductive discussions. Only a change of focus, presenting new methodological tools, will enable significant advances. As mentioned, the National Mapping, despite its shortcomings, is a first step in this direction.

Our argument is as follows: We begin by pointing to issues we believe to be necessary for understanding the realities of the global South, particularly in terms of the validity of the institutional approach of the solidarity economy that is usually accepted in the North. For demonstration purposes, we will highlight the case of the informal solidarity economy (section 1). Next we will assess the state of statistical production in Brazil with regard to the most common forms of organization of the solidarity economy. This will allow us to underline how useful conceptual and methodological developments can be for the objectification of these social practices (section 2). After highlighting the importance of collaborative research on the solidarity economy, we will learn as much as possible from the lessons brought about by the Mapping, at the conceptual and the methodological levels, while also identifying the limits imposed on it in light of the prevailing intellectual outlook. We will conclude with some general considerations aimed at outlining a future system of statistics on the solidarity economy (section 3).

1. Solidarity economy in the global South

The principles of economic solidarity have circulated since the inception of industrial capitalism in the nineteenth century, when they gave shape to the associative, mutualist and cooperative currents of the social economy in many countries from the North and some from the South.[1] From social turmoil among populations facing increasing proletarianization arose solidarism. As the purpose of these various currents was to ensure that all members benefitted fairly from the economic activity, they adopted autonomous and participatory management forms from the very beginning.

In addition, before showing signs of weakening in the first decades of the twentieth century, the social economy opposed the trends of reducing the economy to the market principle and to the rationality of capital accumulation. In that way, it played a considerable role in the construction of the welfare state. As of the 1970s, the crisis of Keynesian regulation and the resulting social imbalances gave rise to a series of new social experiments (Gaiger and Laville, 2009) that reinvigorated associative and solidarity-based economic practices. In particular in Europe and in Quebec, the social economy resumed its critical and participatory impulse, while analogous experiences emerged or revived in the South. In that context, a wide range of formations – such as associations, informal groups, cooperatives, self-management companies, local initiatives in the field of social services and assistance to the needy, social economy enterprises and solidarity-based financing, as well as related mechanisms for promotional and representative organizations, expanded among social categories that had been relegated to the margins of conventional systems of employment or income generation or that had become frustrated in their individual and collective aspirations.

It is thereby justified to speak of, for much of Europe and South America, a new social economy, hereinafter referred to as the social economy or the solidarity economy. As mentioned, the more common concept in South America is that of the solidarity economy, understood to be economic initiatives aimed not only at generating employment and income for its members but also at enhancing their quality of life at large, for example through social benefits or greater public recognition and possibilities for citizen participation. Solidarity concerns cooperation in the productive realm and in the socialization of the means of production,

---

[1] *South* and *North* are metaphors to describe, broadly speaking, the periphery and the center of the global economic and geopolitical system, according to the language of new colonial studies, especially the accounts by Boaventura de Sousa Santos (Souza Santos and Meneses, 2009).

dissolving the separation between capital and labor that is typical of wage employment. For that reason, on the South American continent, the solidarity economy is seen by many authors as an economy that is distinct and separate from the capitalist economy, or else as a future alternative to capitalism. However, despite the many nuances and variations of this kind of economy, the South and the North are increasingly coming to a mutual understanding of the social and/or solidarity economy and are collaborating in efforts to globally align their goals and actions around alter-globalist causes.

Given their common characteristics and their current convergences, these alternative economic experiments concern more or less the same historical process, whereby they are measurable with nearly identical categories of classification and analysis, apart from some minor adjustments to each particular context. That is what emerges from a recent report by the International Labor Organization (ILO) (Fonteneau *et al.*, 2011) that identifies cooperatives, mutual societies, associations and social enterprises as the most common, and thereby representative, forms of the social and solidarity economy at the global level. Other forms tend to be local expressions or informal variations of these reference modalities, such as African *tontines*, seen to correlate to the European microfinance networks (*id.*, p. 2). However, as these variations are generally grouped into one and the same category, usually that of the *other* type, the resulting designations are usually generic, vague and undifferentiated. Usually, in discussions on a specific aspect of the content or on the historical origin of these variants, typically those from the South, the emphasis is on the destitution and vulnerability of the popular classes of these regions, especially in the case of the informal economy (*id.*, p. 14).

Such procedures prevent us from better understanding the realities of the South. In South America, the emergence of the solidarity economy in the 1980s gave continuity and renewed support to a long and rich history of popular solidarism. Throughout the continent, the solidarity economy has remote antecedents, starting with indigenous pre-Columbian forms and collective systems adopted by freed slaves (in Brazil, the *quilombolas*). The varied landscape of the continent includes realities determined by the precarious conditions of wage workers integrated to the peripheral underdevelopment economy, yet also areas in which communities, especially aboriginal peoples, managed to protect their ways of life and keep the capitalist labor market at bay. The latter usually comes at the price of cultural marginalization and extreme poverty.

These various configurations, amplified by national and regional contrasts, render it difficult to draw one overall portrait of the social or solidarity economy. They also explain why distinct manifestations and terms coexist in South America, such as *Popular Solidarity Economy*, *Community Economy*, *Labor Economy*, *Socioeconomy* and *Good Living*, among others, and which span from the informal collective economy to cooperative sectors. If there is a common denominator, it is the inability or refusal of their direct protagonists to live according to the precepts, in terms of their intrinsic sociability, of a society that emerged from peripheral capitalism. In many cases, these manifestations express a refusal to abandon social systems in which economic and social relations are intertwined and in which reciprocity and trust prevail. The solidarity economy essentially aspires to such forms of living and engages to either rescue them or to work towards their establishment (Gaiger, 2008).

That said, the categories of analysis from studies of the North cannot be indiscriminately applied to the South. In most South American countries there was no associative or mutual societies sector, which has its roots in nineteenth century Europe. The only sector present in a reasonable number of South American countries, and especially in Brazil, was that of the cooperative sector, which represents only a fraction of the broader concept of the social economy (Gaiger and Anjos, 2012).[2] This was the case in spite of the fact that community and associative life was and is remarkable in many regions and that self-management was frequently proposed by the labor movement, at least until the advent of populist regimes in the course of the twentieth century, when the state got in charge of the direction of economic and social development. However, it is common to see the solidarity economy of the South being addressed in the legal forms it came to adopt to conform to the canons of modern economics and bureaucratic rationality or, for those who refused these legal forms, remaining in informality. For this angle, one might be prone to interpret the meaning of solidarity economy without consideration of its endogenous propellant principles. That tendency disregards the existence of a historical resistance guided by well determined values and experiences, as has been demonstrated in exemplary studies about the North (Petitclerc, 2007).

We now examine the institutional approach, which subscribes to formalized and institutionalized forms of the social and solidarity economy. It is more appropriate to the realities of the North, which is where such

---

[2]   In South America, the use of the term "social economy" is very inconsistent in that it is applied regardless of whether an institutionalized sector exists (as in Argentina, to some extent) or not. In general, it simply designates earlier forms of the solidarity economy, resulting in the growing preference for the term "social and solidarity economy." Therefore, representative statistical data on the social economy are not available. For a conceptual analysis, focused on the case of Venezuela, see Bastidas-Delgado (2001).

forms originated. Its application in the South is more difficult to assess and evaluate. In general, this requires obtaining a detailed and thorough overview of the actual practices comprising economic solidarity in the South, and then, in a second step, inferring whether institutional formats were adopted either voluntarily by the South or foisted on them; or, whether these formats were indeed rejected by the South, in which case they opted for informality.

The greatest discrepancy between the realities of the North and South is in fact *informality*. In both Africa and Latin America, informality characterizes the popular economy, which is a major focus of the solidarity economy. Contrary to what is implied by the assessment of the ILO mentioned above (Fonteneau *et al.*, 2011), informality is not about *other* cases, but about some of the *most important* cases. Thus, the predominant forms of economic solidarity are characterized by informality and its unique internal logic. By addressing this issue in more detail, this study demonstrates the necessity of refining our conceptualization and reconfiguring the usual categories of analysis before laying the foundation for a system of statistical classification of the solidarity economy in any country of the South.

The history of informality in Latin America is usually considered to span over the last five decades, during which populations migrated from rural areas to urban spaces at a rapidly growing rate. More often than not, the formal labor market in the cities proved incapable of absorbing the majority of people seeking work, nor were there means to ensure their gradual integration into the economy. That contingent of society was thus left to its own devices and forced to subsist on temporary labor. This, in turn, modified the urban landscape, giving rise to peripheral neighborhoods and expanding the informal economy into a phenomenon of great magnitude.

Back then, informality was interpreted either as a marginal residue of capitalism or as a functional element of the reserve army of labor. Such views generally regarded informality as nothing more than a reaction to the modern economy, and as such as a sign of dearth and impotence. Thus, seen to have no agency of their own, these marginal sectors were of no interest to critical theories of peripheral capitalism (Pamplona, 2001; Lopes, 2008). Yet, at the same time, a number of dominant theories and policies focused on the informal sectors, converting these into a target of social assistance programs with a view to their insertion in the market, namely through supporting micro-entrepreneurs (Gaiger and Corrêa, 2010).

Over the years, the spread and persistence of informality led to the belief that it was in fact inserted in popular strategies of resistance and social mobilization. In countries such as Chile, Peru, Brazil and Uruguay, organized movements emerged in urban peripheries (e.g., the *favelados* and *pobladores*) which fought for housing, urban services, income and employment (Bell Lara, 1997). Community initiatives multiplied and gradually aroused the interest of civil organizations, churches and development institutions, particularly microcredit institutions. Most of these initiatives started their operations through the pioneering women's banks and gave rise to grassroots communities, neighborhood associations, unions of family growers and, already in the 1980s, the first collective experiences of income generation that were the precursors of the solidarity economy.

Informality was then reinterpreted as being part of the so-called popular economy, which had its own social logic of promoting community ties and associativism. This new vision thus offset the poor light in which informality was cast in previous theories.[3] Coraggio (1999), for example, considers the popular economy to have a rationality of its own that is guided towards the formation of a collective labor fund, namely through individual and collective strategies that are inseparable from the mesh of social relations in which small-scale economic agents act. The effectiveness of such strategies, then, is seen to depend on the relative freedom prompted by informality. Thus, the material and social assets typical of the informal economy should not be underestimated but rather valued by means of social emancipation projects.

In 2007, *informal groups* accounted for 36.5% of all enterprises surveyed by the first National Mapping. Several of them have prospered while retaining their informal traits (Gaiger, 2011). In such cases, workers shed an attitude of constant adaptation to circumstances in favor of actions allowing them to take more control of the productive factors and of predicting future consequences. In other words, they then see themselves as a force capable of creating new situations and of influencing the transformations they seek. This change is favored when workers rely on their social relations. A metamorphosis turns personal ties and tradition into a properly enterprising and solidarity economy logic that is sustained by cooperative relationships (Razeto, 1990). Once equipped with such a new foundation, enterprises can contribute to overcoming instability and the uncertainty affecting the material life of the poor, because they

---

[3] This optimism sometimes turned into idealism, in its tendency to separate the informal and the popular economy into two worlds and to focus only on combative community expressions of the popular economy, disregarding the coexistence of different principles of value and the most diverse arrangements to ensure survival, including despotic practices that foster individualism and inequality (Gaiger, 2009). In any case, comprehensive reviews of literature on informality, as in Lopes (2008), fail to find a consistent consensus on the theme.

attenuate their subordination to the dictates of the economy and redistribute a greater portion of the surplus value to the workers (Gaiger, 2006). From the viewpoint of economic culture, such enterprises contribute to the rationalization of solidarity since they stimulate intentional and everyday practices of solidarity.

Thus, even if it lacks an appropriate institutional framework, informality is a part of institutionalized forms of economy insofar as it complies with rules that determine the management of informal businesses. *Informal, popular* and *solidarity-based* are not equivalent but compatible terms. In the South, their overlapping compels us to see the logic of informality as an attribute underlying most enterprises of the solidarity economy and as a basis of current economic conditions.

Moreover, the presence of institutional forms such as associations, cooperatives mutual societies and other social economy enterprises do not have the same implications in the North and the South. More specifically, in the South, their presence is less indicative of the realization of the principles of the solidarity economy than in the North. Overall, the quest for economic solidarity has different motivating forces in the North and the South, and the resulting legal forms and structures which these efforts manifest in cannot be taken as synonyms or correlates, even if they bear the same name. Clarifying this confusion between the terminology between the South and the North would involve determining the meaning of economic solidarity in the South, followed by a definition of the criteria, indicators and institutional arrangements in which the spirit of solidarity is manifested.

2. A conceptual and statistical hiatus

How do Brazilian official statistics refer to the most common forms of organization in the solidarity economy? Treating these forms separately will shed light on the inadequacy of statistical data. Moreover, it will show the usefulness of clarifying such realities in order to identify what emerges genuinely out of solidarity and its respective economic rationality.

These aforementioned conceptual problems have prevented studies from becoming interested in the phenomenon of solidarity in the informal economy.[4] In addition to these problems, which generally give rise

---

[4] One of our studies on this subject arrived at some enlightening conclusions: Solidarity provides networks and external support to informal enterprises, breaking up their isolation and increasing their chances of survival. By being enterprises amid people who gather under equal conditions and who made a rational choice for self-management, these forms of the solidarity economy stimulate the participation of their members and,

to negative views on informality, there is a lack of minimally systematic statistics about the informal economy. Finally, as any available information relies on conventional economic categories, relying on it may be a trap even if it can provide useful insights.

In Brazil, the IBGE (Brazilian Institute of Geography and Statistics) is the leading government body in charge of performing research and statistical analyses, including demographic and economic censuses. To date, the IBGE conducted only two national research projects on the informal economy, in 1997 and 2003 (IBGE, 2005), providing data for major quantitative studies on this sector since then.[5] As part of those projects, non-agricultural work and production units that are destined to the market (vs. food self-sufficiency) and that are owned by people who work either for themselves or with family members, unpaid employees or up to five employees have been classified in the category of "informal enterprises." The absence of legal formalization, in turn, was considered secondary since the units are characterized by a low gap between capital and labor as production factors (IBGE, 2005). Therefore, wage labor would not be their base of operation, nor should the rate of profit be considered as a key variable. The main criterion of informality is not the lack of legal personality but its particular dynamic in which the company's capital and its physical agents are inseparable.[6] Economic management should not be set apart from the needs of its members, since without them the business loses its meaning and mainstay: labor.

However, the dynamic of the informal economy is usually attributed to the tiny scale of its economic operations and to the family-based nature

---

unlike informal microenterprises, aim for the equitable distribution of gains (Gaiger, 2011).

[5] "It must be understood that the sector classification [used by IBGE] regards the form of organization of production units (enterprise approach) whereas the employment concept refers to the characteristics of job links (labor approach)" (Hallak Neto *et al.*, 2009: 15). The second one prevails largely in the Brazilian literature on informality.

[6] From the point of view of the Brazilian System of National Accounts, in compliance with the United Nations "the informal sector can be understood as a subdivision of the Households institutional sector in which are classified the non-agricultural production units characterized by a low level of organization and for not having a clear division between labor and capital as production factors and production of which is primarily designed for the market (...). The Households institutional sector includes units of consumption and production, being defined as a small group of persons who share the same living accommodation, who pool some, or all, of their income and wealth and who consume certain types of goods and services collectively, mainly housing and food. This sector includes production units consisting of self-employed workers and employers of unincorporated enterprises. The expression 'unincorporated enterprises' underlines the fact that the production unit is not an independent legal entity separated from the household members who own them" (Hallak Neto *et al.*, 2009: 8-9).

of business operations, both of which are usually associated with inefficiency (Pamplona, 2001; Lautier, 2004). The criterion adopted by the IBGE coincides with the peculiar trait of solidarity enterprises of operating simultaneously as economic societies and as "people societies," without separating the enterprise's capital from its membership. More than simply a matter of low complexity or amateur management, the rationale behind the criterion is consistent with the facts. Understood in these terms, this trait pertaining to the *spirit* of informal economy should not be lost but preserved by formalized modalities of the solidarity economy (Hespanha, 2010; Gaiger, 2012d).

Yet, from this same *spirit* arises the logical thread that also explains why the adherence of informal solidarity enterprises to available formal alternatives is conditional and problematic. Entrepreneurs of such enterprises are generally not interested in a legal framework that requires them to adopt an economic behavior determined primarily by the efficiency and viability of their enterprise at the expense of human relationships, solidarity and well-being. Nor are they interested in committing to obligations, be it with regard to the use of their time or resources, that are beyond their means and preferences. Finally, most are not ready to give up the *freedom* of informality in exchange for an efficiency, achieved through technical knowledge and bureaucratic rationality, that has historically subjugated them.

We advance the hypothesis that, the unique circumstances of each historical context aside, this understanding is what has led workers, producers and consumers to adopt the *association* and the *cooperative* as the only institutional alternatives available in Brazil for preserving trust and cooperation in legally recognized collective economic activities. However, by opting for these alternatives, players are invariably forced to deal with their respective drawbacks. Let us consider the case of associations and then that of the cooperatives.

One of the few studies on this subject examined the resistance of urban recyclable garbage collectors to organize themselves in cooperatives and their preference for associations, which are more flexible and adjustable to their individual interests and needs (Souza, 2005). In fact, the *association* is the most widely used juridical status in the solidarity economy: in the first Mapping, 52% of enterprises were associations; in the second, 59.9%. As the percentage of informal enterprises declined simultaneously in the same proportion, it is plausible to suppose, hypothetically, that the association works as a preferred alternative for those who decide to leave informality. However, despite being a reasonable option, it is still an incomplete solution for several important reasons.

What does an association correspond to, in juridical terms? In Brazil, the legal framework of associations is extremely broad, because it encompasses groups of people who engage in activities that do not necessarily have an economic purpose and that are not affiliated with the work of more established institutions such as churches, foundations and political parties. The activities of associations may involve waged professionals and generate economic dividends as long as they meet the social aims of associative entities and are not used for the private enrichment of its members. Yet, associations may subcontract work to organizations or create subsidiaries that, for their part, are not bound to the same criteria and that be profitable. In this way, the legal framework of associations leaves the door open to a number of ambiguities and contradictions. Nevertheless, it contains an element of great interest to the solidarity economy: decision-making must be exercised on an equal basis by the members, with no interference of their individual quotas of capital in the associative entity.

Community clusters in urban peripheries, cultural and leisure centers, and big professional sports associations such as soccer clubs are considered to be associations. Given the impossibility of regulating such a wide and heterogeneous range of entities, associations are essentially subordinated to the requirements of the sector of activity in which they operate. Analogously to the Third Sector in Brazil, associations are defined more by what they are *not* (or cannot be) and less by what would characterize and unify them (Fernandes, 1994; Gaiger, 2012a). Overall, it is a sector with no encompassing social identity and no oversight bodies that keep statistical records.[7]

There are many stories behind associations, among them, popular associativism. Since the 1970s, in the context of demographic flows that resulted in current urban areas, the association has been a popular instrument of organization and struggle for the right to housing and for decent living conditions. In countries such as Brazil and Chile, the role played by community organizations is visible as a pillar of broader social movements, such as the democratic struggles and electoral clashes that led to the renewal of political parties and shifted governments towards the left. Simultaneously, associations functioned as the driver of local initiatives, giving them impulse and institutional backing. As a result, community

---

[7] The statistics only cover the entrepreneurial foundations and NPOs, whose sum equals roughly the Third Sector in Brazil. The latest study on this subject (IBGE, 2012) in 2010 reported over 290 thousand institutions of this kind, comprising 52.2% of all nonprofit organizations registered at the General Register of Brazilian Companies (CEMPRE). Yet, such statistics run into the same difficulty: it is not productive to account for and compare such disparate things as community initiatives, NGOs, philanthropic associations, foundations and other entities only on the basis of their being either private or non-profit oriented.

projects for generating income and economic development, when benefitting from the juridical support of associations, are often intertwined with the latter. The result is a hybrid: social community activities that essentially pursue economic enterprise aims and that, as such, operate in a legal grey zone. The most common compromise is to use the legal incorporation of the association, to prevent full informality, and to postpone the formalization of the enterprise, thereby foregoing the privileges and benefits conferred upon the lawful exercise of economic activities.[8]

A similar scenario has long characterized the rural world of the solidarity economy. Usually, small farmers associations are hired by collective enterprises to perform production, trading or service activities.[9] When legally indispensable, they use the personal registration of their members, employed in their individual business as family growers. Given their generic status, associations can pursue a number of economic strategies. Taking the status of an association is in a way a form of subterfuge. It grants institutional recognition to solidarity enterprises so that these can operate in a semi-formal way and receive support and subsidies. However, as shown by some studies (Pinto, 2006), associativism is a *spirit* that goes beyond pragmatism. It constitutes a collective path in which identities and solidarity practices were forged and now revalued. Therefore, in rural and urban areas, solidarity enterprises in general fit into broader collective structures, allowing them to overcome the abandonment and isolation experienced by many micro and small enterprises in Brazil.[10] Separating their economic goals from their social purpose would be contrived (Gaiger, 2011; 2012d).

To ensure that such links remain intact, the third option sought by solidarity enterprises is the *cooperative*. Politically, the solidarity economy has questioned official cooperativism in Brazil due to the inconsistency between the doctrinal principles it is supposed to defend and the historical

---

[8] The informal economy basically depends on the compliance of public authorities to operate. It lacks access to credit and to environmental or sanitary permits indispensable for certain activities, which are subsequently forced to function clandestinely.

[9] In Brazil, rural associativism has played a vital role for small family farmers ever since they occupied agricultural areas in the course of the 19th century. Yet, by and large, rural associativism has had no distinct image of its own and no representative mechanisms. Besides, it was largely co-opted or stimulated to act as a cog in the wheel for oligarchic domination. In the interior of the country, it was common practice to create or favor associations in return for the loyalty of their members to the lords – strongmen and oligarchs – from amongst whom political power derived and ruling elites emerged.

[10] For example: the lack of policies and support programs for micro and small enterprises; scarce collaboration practices among them as well as restricted associative networks; and higher sensitivity on their part to the competition caused by the increased number of companies (driven by the "necessity entrepreneurship").

development of the sector in the country, especially with regard to its legal aspects.[11] A new model would be expected to supersede traditional cooperativism and restore the inherent trust in cooperatives. The latter have lost credibility due either to their lack of internal democracy or to having been fraudulently created by other enterprises as outsourcing businesses to evade social costs.

The proliferation of solidarity cooperatives, like that of associations, is understandable from a pragmatic point of view. Although there is a striking gap between the Brazilian cooperative statute and the aspirations of the solidarity economy, the legislation fails to provide adequate alternatives for the incorporation of companies formed from the free membership of people wishing to cooperate on an egalitarian basis in activities with economic goals that are nevertheless not aimed at profit. Self-managed companies, credit societies and productive or service enterprises can only establish themselves legally through their formalization as cooperatives. Moreover, they need to comply with a complex and usually non-specific set of complementary laws that do not distinguish cooperatives from other companies and that subject them to the harmful effects of this isomorphism.

For the most part, the cooperatives registered in the first Mapping had started their activities in the last 15 years, convinced that this institutional form is, *in theory*, a valid alternative to meet the interests of workers who opted for self-management and economic solidarity. Given the gap of several years that lies between the mappings, the two mappings should capture important changes in the number of cooperatives, despite the inaccuracies arising from the *snowball* methodology, mentioned below. However, the percentage of cooperatives declined rather than increased between the first and second Mapping, namely from 9.7% to 8.8%. Albeit only slightly and perhaps reflecting methodological flaws of the mappings, this decline goes counter to an expected increase. The latter was anticipated since solidarity economy programs in Brazil focused primarily on the creation and strengthening of cooperatives. Once

---

[11] The national cooperative legislation was established at the time of the military regime, more precisely in 1971. It sought to accommodate the interests of the cooperative sector, which since then has been in control of the Organization of Brazilian Cooperatives (OCB), the only national body recognized by the clause of representative unity. The law is too generic to promote and regulate cooperatives, but also imposes bureaucratic requirements that hinder the formalization of solidarity enterprises, for example by setting the minimum number of members required to found a cooperative at twenty. Due to its historical origins and political profile, the OCB lacks legitimacy to attract new sectors, which explains the emergence of independent currents, such as the Confederation of Agrarian Reform Cooperatives in Brazil (CONCRAB), which is linked to the Movement of Landless Rural Workers.

again, the drawbacks seem to outweigh the attractiveness of this third alternative to informality, institutionalized for a long time but permanently controversial.

Cooperativism was originally introduced to Brazil by European immigrants, in the late 19th century, as a way to overcome situations of extreme poverty and destitution. In these early days, consumer cooperatives emerged, and so did credit and farming cooperatives, especially in the south of the country. Consumer cooperatives expanded in the 1950s and 1960s. Subsequently, urban cooperativism showed signs of stagnation, resulting in a number of barriers to its expansion and survival. In turn, farming cooperatives were gradually encouraged, primarily by military governments who sought to boost exports by means of expanded agricultural yields. Since 1970, the prevailing farming cooperativism in Brazil has strengthened the dominance of the conservative elite, traditionally focused on the agribusiness export economy, and has served as a political-corporate alliance extremely sensitive to economic power. This explains the strong dependence of farming cooperativism on government policies and on the skills of its leaders to negotiate and deal with the state.

This framework was supported by a policy of social control and state intervention that brought no significant change to cooperative workers in rural areas. On the contrary, the model spread distrust about cooperativism amidst small farmers, who had decades prior valued this instrument of economic development and community strengthening. Meanwhile, the urban movement was given a new impetus with the creation of many worker cooperatives in the 1980s. At the height of the proliferation of such cooperatives, several studies indicated that they were utilized mainly to make working relations more flexible, to outsource services and to cut labor costs (Lima, 2007). Nevertheless, genuine cooperatives were also identified, such as *reclaimed factories*, one of the first divisions of the solidarity economy.

Today major cooperatives function as capital companies, aimed at profitability in the market and engaged in professional and efficient management. At the other extreme, small cooperatives in urban peripheries, focused on the socio-economic inclusion and basic needs of poor populations, have an egalitarian nature, appreciate the fact of governing themselves collectively, and identify themselves with the solidarity economy (Nunes, 2001; Anjos, 2012). By their side, there are false cooperatives, which use the legal framework to exploit manpower at low cost but that preserve the hierarchy as well as the social division between capital and labor in the enterprise (Leite and Georges, 2012; Gaiger and Anjos, 2012). Therefore, the Brazilian cooperative sector is heterogeneous with regard to the nature and scale of its activities, the complexity of its cooperative organizations and, fundamentally, its ideological principles.

Statistical data on cooperatives are available, since all economic enterprises must provide periodic reports to the supervisory body responsible for their field of activities, such as labor, credit, health or trade. These records, being specific, could be useful for analyzing concrete issues, such as the hiring and dismissal of personnel, fluctuations in membership and volume of commercial activity. However, extensive statistical series, such as the Brazilian Economic Census, treat cooperatives like other companies and offer very sparse and irregular specific information, especially in the case of agricultural cooperatives. Besides these administrative records, the OCB has information on the various branches of Brazilian cooperatives, particularly on the evolution of membership, jobs and major economic figures. But, because such data are provided to the OCB on a non-compulsory basis, they do not encompass the totality of existing cooperatives and are subject to omissions, errors and discontinuities. Furthermore, like official records, they do not allow for the discrimination of the nature of cooperative practices or of its democratic and solidarity-based character.[12]

In conclusion, the currents that characterized the social economy in the North also appeared in Brazil, but as episodic and less comprehensive social experiences. In the few times they acquired considerable weight, such as with cooperativism, they faced great obstacles in order to maintain their own profile and their role as an alternative to the prevailing forms of the economy. Instead, their role was relegated to that of an auxiliary force or compensation device of the social costs of capitalistic economic development. However, the advent of the solidarity economy, as a counter-current, originated largely from other, prior forms of solidarity that are little known and poorly recognized in their value. Overall, it constitutes one of many new modes of collective action that emerged in the last twenty years, sometimes under the influence of international initiatives, such as fair trade. Its development took place without clearly defined conceptual frameworks, appropriate indicators, harmonious regulatory frameworks and representative statistics. The National Mapping has not solved these problems; however, it does point them out. Herein lies its value.

---

[12] As an example, the proliferation of *false* labor cooperatives, as mentioned above, has greatly inflated the figures of this sector and sparked widespread criticism, also feeding distrust of cooperatives themselves. As statistics cannot distinguish between genuine and false cooperatives, that mission has fallen upon regulatory agencies, which also led to complaints about excessive rigor.

3. Contributions and limitations of the National Mapping

In Brazil the first studies on the solidarity economy were sponsored and promoted by an entity that supports popular movements. The results were first published in newsletters, magazines and books. Only later did the subject become of interest to academia, in particular to research institutions who were already devoted to the study of associations and cooperatives. With time, universities began developing popular cooperative incubators and social projects in this field. Since these beginnings, a close and often long-term interaction between reflection and action has marked the creation of knowledge about the solidarity economy, with scientific activity taking place even beyond the walls of universities. Researchers and professors working on this theme are usually engaged in practice-oriented support programs and also participate in public debates and discussions. The research projects develop in conjunction with demands from social actors and are linked to public policies that fund studies and assessments. Overall, the academic field working on this topic is notable for intense exchange between civil society, political institutions and the state.

This collaborative practice had its greatest exponent in the two editions of the National Mapping. Once the available methodological alternatives were evaluated, the mapping was planned to be carried out as a process for mobilizing actors of the solidarity economy, with support of research institutions and the government. The goal was to broaden as much as possible the scope of the data collection, through successive identification of enterprises made by the enterprises already researched (*snowball* effect) and, above all, through a commitment of everyone involved to contribute to the discovery and recognition of the least known realities that had been undervalued and poorly integrated in the organized sectors of the solidarity economy. Remote places of the country were to be reached, converting the protagonists of these experiments into visible actors. In order to enter rural and remote areas of the national territory, 230 entities and hundreds of interviewers were engaged and trained to participate in the first Mapping, a research that went on for almost three years. In the same period, the solidarity economy underwent moments of great expansion and held its largest meetings in the country. Although it achieved only an incomplete survey of the nation's reality, the bottom line showed that its chosen strategy was on par with conventional research methods, given the cost-benefit ratio and the fact that only a collaborative research including its own participants and partners, supporters and networks would ensure new knowledge on the solidarity economy.[13]

---

[13]   The long duration of the first Mapping created problems for counting and data standardization. Its greatest value lies not in scientific rigor but in the fact that it was the first

Until the Mapping, the lack of systematized and representative data on the solidarity economy restricted empirical research in Brazil to a mostly qualitative approach, usually through case studies. Empirical comprehensive analyses allowing to identify structural trends and significant changes in the profiles of the enterprises were carried out only on rare occasions. As a result, many theses on the solidarity economy were limited to apriorisms or conjectures, devoid of empirical foundation and factually sustainable theoretical premises. The Mappings, however, opened the doors to overcoming such limitations (Miranda, 2011; Anjos, 2012; Gaiger, 2012b).

Their first valuable contribution was to have proved the existence of a wide range of enterprises in which solidarity is a core ethical value and crucial component in the administration of business operations. In other words, these enterprises were found to ensure their efficiency and viability by subscribing to principles of self-management, cooperation, equity and the pursuit of well-being for all members, rather than seeing their values come in contradiction with the enterprise's efficiency. The main analysis on the first Mapping (Gaiger, 2007) sought to put to the test and develop an ideal-typical concept (in the Weberian sense) of the solidarity economic enterprise (SEE) and to, moreover, examine precisely how the combination of entrepreneurship and solidarism is able to convert such organizations into a unique social and economic whole.[14] In general, it was noted that solidarity enterprises, with a few exceptions, cultivate a set of practices guided by a rationality that combines solidarity and economic performance. Moreover, more solidarism was generally found to lead to improved entrepreneurship. Although this rule does not apply uniformly, it does account for the degree of success and survival of a significant number the enterprises. This, in turn, offers decisive support to theories that interpret such practices as expressions of *another* economy.

The Mapping project questions preconceived ideas. By confirming the existence of a considerable number of organizations run by

---

comprehensive national survey. Indeed, it was conducted in 2,274 municipalities in the 27 Units of the Federation (states) that make up Brazil. The collected information relates to the initial conditions of the enterprises, their development strategies and the benefits provided to their members and social environments. The conceptual basis and methodology of the first Mapping can be found on www.sies.mte.gov.br.

[14]   The concept is based on the assumption that solidarity economy organizations, despite their diversity, tend to fit into a logic that gives them some level of accumulation and growth, providing them with stability and feasibility, yet with the particularity that they trigger, to this end, an economic rationality grounded on members' involvement in management and cooperative work. The analysis was based on a multifactorial method and culminated in an assessment of the degree to which the *entrepreneurial vector* and the *solidarity vector* are represented in solidarity enterprises.

employees who cooperate with one another, it shows that the solidarity economy is alive and real and that it is distinct from mere informality, subservient associativism and false cooperatives. In addition, certain data of the Mapping gave way to surprise. For example, women were shown to work proportionally more than men yet also to assume leadership roles in a considerable percentage of enterprises, which were prominent for their community and social involvement. A further incongruity is the remarkable concentration of enterprises in the countryside of northeastern Brazil. The latter is explained, in part, by the prevalence in those regions of a popular autochthonous associativism that is disconnected from political domination structures. This topic has been examined in studies on *coronelismo* and oligarchic clientelism (Singlemann, 1975; Domingos and Hallewell, 2004). The Mapping data demonstrates the presence of an associative current ignored by prevailing interpretations, which tend to argue that associations and cooperatives originated primarily in the south, in wake of immigration waves from Europe that began in the 19th century.

To unveil *new* realities – in the eyes of intellectuals and the very actors of the solidarity economy – the essential value of the Mapping is of an epistemological order. Although producing only a partial portrait of the solidarity economy, the Mapping was an exceptional effort to expand horizons, transcend social boundaries and achieve recognition for realms that had been condemned or discredited before. In these *lost* places of the rural interior of the country or urban neighborhoods, populations live in peripheral social and economic circuits that are insignificant from the perspective of the modern capitalist economy but that are invaluable as repositories of ways of living in solidarity and reciprocity. Embedded in popular culture, such primary forms of social cohesion fulfill functions that are indispensable for the survival and integrity of the human populations that cultivate them. They are more than simple anachronisms or incomplete versions of newer types of solidarity that have emerged and that are cast aside as yet another sub-branch of the solidarity economy.

Thus, *new* categories of social actors – although they are actually ancestors – have entered the solidarity economy arena and have begun to appear on forums and representative bodies. Small-scale fishermen, *Quilombolas*, indigenous peoples, rubber tappers and other traditional segments of the population got on the scene, multiplied the number of organizations of the solidarity economy and pushed for innovations in public policies. The small-scale fishermen, for example, organized to form the Fisheries Solidarity Network, one of the oldest networks identified by the Mapping. Moreover, specific social categories (women, beneficiaries of social programs, youth at risk, and people with a disability or mental disorder) have been increasingly considered in public policies since the Mapping and are now have a greater presence in the solidarity economy.

This progressive social enlargement was better recorded by the second Mapping, which contains additional variables to characterize the population involved in solidarity enterprises.[15] But, this enlargement also gave reason to believe that the current boundaries of the economic sphere are arbitrary and inappropriate, as advanced by the seminal thought of Karl Polanyi and the New Economic Sociology. A topic of continual debate among the social actors of the solidarity economy, it in fact caused methodological deadlocks in the Mapping. In solidarity enterprises, several activities transcend the economic sphere and respond to immaterial needs and aspirations that are of a social or cultural nature. It can therefore be difficult to identify the priority goals and social profiles of the participants of each solidarity enterprises. Moreover, the enterprises have structural and functional complexity, as they combine collective and individual activities, permanent or transitory, whose variable importance results in shifting forms of involvement of the members. With such fluidity and inaccurate contours, the enterprises and their activities can be difficult to fit into classification schemes.

When encompassing this diversity, encrypting it and encouraging comparisons, the Mapping provides raw material for reassessing our conceptual tools. Momentarily, it allows us to put aside the institutionalized forms established by solidarity enterprises, since such choices fail to reliably reflect their nature and to recognize the existence of open social processes. At present, there prevails a push toward institutionalization, which must be considered in any analysis of organizational solutions adopted by enterprises, when faced to opt for one of the already settled options of the economy. A closer examination of the practices of the enterprises, as recorded in the Mapping, is a promising path. However, preliminarily, this requires overcoming some obstacles, a legacy of the Mapping and certain intellectual habits. In the following, we shall examine this challenge in more detail.

In order to identify its target population, the Mapping selected a set of criteria that would cover the greatest number of *potentially* solidarity-based economic organizations. Such organizations were to have a permanent economic purpose, be under the ownership or control of members who belong to more than one family unit and who themselves perform

---

[15] The second Mapping contains a larger amount of information on the educational, professional and ethnical aspects of the population engaged in solidarity enterprises, among others. Data will be released on the website sies.ecosol.org.br/.

the target activity (production, services, trading, consumption), only occasionally employing non-member workers and, if so, through collective
management systems. Although any organization complying with these
minimum criteria were to be included in the target population, several
teams of the Mapping stipulated *ad hoc* restrictive criteria by way of re-
interpreting criteria or adding complementary requirements. An example
of this is the exclusion of a number of enterprises that employed workers from outside their membership base. In turn, this generated debate
about the appropriate and acceptable kinds of labor relations for solidarity enterprises, with the need to arbitrate between members and the
needed skilled workers in multi-skilled organizations or with fluctuat-
ing demands of workforce. Thus, combined with the understanding that
veritable self-management is irreconcilable with the social division between capital and labor, these criteria stripped the Mapping of a number
of organizations commonly classified within the social (and solidarity)
economy in other countries. Furthermore, they eliminated a significant
number of cooperatives suspected of belonging to the capitalist economy, of hiring paid labor or for having inadequate business management
models.

None of the criteria allowed to consider and include hybrid cases,
in other words, to assess enterprises that corresponded only in some re-
spects with the solidarity economy and to include these as peripheral,
incomplete or circumstantial variants. The biggest barrier, however, was
the tendency of social players and mediating agents to classify realities in the binary terms of either inclusion or exclusion. This may have,
as in the example mentioned above, placed the solidarity economy in
radical opposition to, the capitalist universe of waged relations. In terms
of ideas, there is an antagonistic dispute between two projects of society, two totalities. Besides, the teleological blemish of each of these
extremes induces to examine real practices under the light of idealization
(Edelwein, 2009).

The choice of explanatory frameworks with such features generally
leads to a somewhat demanding and heavy apriorism that does not necessarily correspond with reality.[16] In Brazil, this phenomenon is observable
in specialized circles, where, discussions on solidarity economy indicators (Kraychette and Carvalho, 2012) are often dominated by a somewhat
dogmatic normative approach concerning the perspectives of social transformation in the solidarity economy. In that context, a clear recognition
of facts becomes dependent on the ideological assumptions that underlie

---

[16] This subject is pointed out especially by Santos (1999; 2004). For a particular discussion on the Brazilian case, see Gaiger (2012c).

such views and their respective political strategies. While this may be understandable in the debates held in the public sphere and in social movements, it is insufficient for the objectivization of reality and its theoretical and conceptual treatment.

Final considerations

Brazil lacks a systematic reflection on the problems identified above, except with regard to the inadequacy of juridical frameworks, indexes and statistics (Wautiez *et al.*, 2003). The creation of a legal framework, insistently called for, would facilitate the delineation of this field of study, as seen in national statistics and concepts employed in the North (Bouchard, 2008; Bouchard *et al.*, 2011). Furthermore, it would avoid the trivialization of the concept of the solidarity economy and its instrumentalization by economic sectors. However, the main inadequacy is the lack of a clear starting point for the construction of the desired regulatory framework. As the solidarity economy is not merely economic, the task would inevitably become complicated and also unfeasible in cases where the peculiar rationality of solidarity enterprises is unknown.

Thus, advancing in such matters requires a comprehensive analysis of the nature of the solidarity economy, followed by an examination of the organizational solutions adopted and the impact of such choices on enterprises. Instead of putting juridical criteria above substantive criteria when assessing the rationale of such initiatives, the starting basis should be the sui generis nature of the solidarity economy. Once a theoretical model is confirmed through a reasonable number of cases, the next step is to investigate whether the enterprises under analysis fit into preexisting formats or, when not, how they provide mixed solutions, such as the juridical figure of the association.

The recommended methodology consists of exploring representative databases that encompass the significant variations of solidarity-based practices and to then proceed with their quantification and comparison. The investigation of regularities and structural variations could allow to advance the effort of building typological constructs that translate the previously identified variants and improve the qualification and classification of enterprises. Developing criteria for differentiating projects would contribute to ulterior studies, both quantitative and qualitative, be it to grasp specific situations, to improve comparability with other classificatory systems or to enhance both the typology and the proposed indicators. Finally, focusing on the theoretical and methodological aspects of this task would allow to render the resulting proposals more realistic, feasible and effective in that it would avert a distinct apriorism or bias towards prolepsis.

References

Anjos, E., *Práticas e sentidos da economia solidária. Um estudo a partir das cooperativas de trabalho*, Doctoral thesis in Social Sciences, São Leopoldo, Unisinos, 2012.

Bastidas-Delgado, O., "Economía social y economía solidaria: intento de definición," *Cayapa – Revista Venezolana de Economía Social*, Vol. 1, No. 1, 2001, http://www.saber.ula.ve/handle/123456789/18604.

Bell Lara, J., "Informalisation et nouveaux agents économiques: le cas de l'Amérique Latine," *Alternatives Sud*, Vol. 4, No. 2, 1997, pp. 19-39.

Bouchard, M. J. (ed.), *Portrait statistique de l'économie sociale de la région de Montréal*, Montreal, Canada Research Chair on the Social Economy, 2008.

Bouchard, M. J., Cruz Filho, P. and St-Denis, M., *Cadre conceptuel pour définir la population statistique de l'économie sociale au Québec*, Cahiers de la Chaire de Recherche du Canada en Économie Sociale, R-2011-01, Montreal, Canada Research Chair on the Social Economy / CRISES, 2011.

Coraggio, J., *Política social y economía del trabajo*, Madrid, Miño y Dávila Editores, 1999.

Domingos, M. and Hallewell, L., "The Powerful in the Outback of the Brazilian Northeast," *Latin American Perspectives*, vol. 31, No. 3, 2004, pp. 94-111.

Edelwein, K., *Economia solidária: a produção dos sujeitos (des) necessários*, Doctoral thesis in Social Work, Porto Alegre, PUCRS, 2009.

Fernandes, R., *Privado, porém público: o Terceiro Setor na América Latina*, Rio de Janeiro, Relume-Dumará, 1994.

Fonteneau, B. *et al.*, *Economía Social y Solidaria: nuestro camino común hacia el trabajo decente*, Turin, International Training Center of the ILO, 2011.

Gaiger, L., "A racionalidade dos formatos produtivos autogestionários," *Sociedade & Estado Magazine*, Vol. 21, No. 2, 2006, pp. 513-44.

——, "A outra racionalidade da economia solidária. Conclusões do Primeiro Mapeamento Nacional no Brasil," *Revista Crítica de Ciências Sociais*, Vol. 79, 2007, pp. 57-77.

——, "A economia solidária e o valor das relações sociais vinculantes," *Katálysis Magazine*, Florianópolis, UFSC, 2008, Vol. 11, No. 1, pp. 11-19.

——, "Antecedentes e expressões atuais da Economia Solidária," *Revista Crítica de Ciências Sociais*, 2009, Vol. 84, pp. 81-99.

——, "Relações entre equidade e viabilidade nos empreendimentos solidários," *Lua Nova Magazine*, 2011, Vol. 83, pp. 79-109.

——, *From the Popular Economy to the Third-Sector; Origins and Buoyancy Forces of the Solidarity Economy in Latin America*, VII Congreso Internacional Rulescoop, Valencia, Universitat de València, 2012a, http://www.congresorulescoop2012.es/comunicaciones/?search-by=autor&search-tema=0&search-keyword=0&search-string=gaiger.

——, "La présence politique de l'économie solidaire. Considérations à partir de la première cartographie nationale," in Georges, I. and Leite, M. (eds.), *Les nouvelles configurations du travail et l'Économie sociale et solidaire au Brésil*, Paris, L'Harmattan, 2012b, pp. 231-258.

——, "Avances y límites en la producción de conocimientos sobre la Economía Solidaria en Brasil," in Coraggio, J.L. (ed.), *Conocimiento y Políticas Públicas de Economía Social y Solidaria, Problemas y Propuestas*, Quito, Instituto de Altos Estudios Nacionales (IAEN, la Universidad de postgrado del Estado), 2012c, pp. 55-84.

——, *The Uniqueness of Solidarity Entrepreneurship in the Fight against Social Exclusion*, Second ISA Forum of Sociology, Buenos Aires, 2012d, http://isarc10internetforum.wikispaces.com/ISA+2012+Session+13.

Gaiger, L. and Anjos, E., "Solidarity Economy in Brazil: The Relevance of Cooperatives for the Historic Emancipation of Workers," in Piñero, C. (ed.), *Cooperatives and Socialism: A View from Cuba*, Hampshire, Palgrave Macmillan, 2012, pp. 212-234.

Gaiger, L. and Corrêa, A., "O microempreendedorismo em questão; elementos para um modelo alternative," *Política & Sociedade – Revista de Sociologia Política*, 2010, Vol. 9, No. 17, pp. 205-230.

Gaiger, L. and Laville, J.-L., "Economia solidária," in Cattani, A. *et al.* (eds.), *Dicionári internacional da outra economia*, Coimbra, Almedina, 2009, pp. 162-168.

Hallak Neto, J., Namir, K. and Kozovits, L., *Sector and Informal Employment in Brazil*, Special IARIW-SAIM Conference on Measuring the Informal Economy in Developing Countries Kathmandu, Nepal, 2009, http://www.iariw.org.

Hespanha, P., "From the Expansion of the Market to the Metamorphosis of Popular Economies," *RCCS Annual Review* (an Online Journal for the Social Sciences and Humanities), No. 2, 2010, http://rccsar.revues.org/210#entries.

IBGE (Instituto Brasileiro de Geografia e Estatística), Economia informal urbana 2003, 2005, http://www.ibge.gov.br/home/estatistica/economia/ecinf/2003.

——, *As fundações privadas e associações sem fins lucrativos no Brasil 2010*, Rio de Janeiro, IBGE, 2012.

Kraychete, G. and Carvalho, P., *Economia popular solidária; indicadores para a sustentabilidade*. Porto Alegre, Tomo Editorial, 2012.

Lautier, B., *L'Économie informelle dans le Tiers-Monde*, Paris, La Découverte, 2004.

Leite, M. and Georges, I. (eds.), *Les nouvelles configurations du travail et l'économie sociale et solidaire au Brésil*, Paris, L'Harmattan, 2012.

Lima, J., "Workers' Cooperatives in Brazil: Autonomy vs Precariousness," *Economic and Industrial Democracy*, 2007, Vol. 28, No. 4, pp. 589-621.

Lopes, E., "Informalidade: um debate sobre seus distintos usos e significados," *News and Reference Bulletin*, 2008, Vol. 65, No. 1, pp. 49-70.

Miranda, D., *A democracia dialógica: uma análise das iniciativas da Economia Solidária*, Doctoral thesis in Social Sciences, São Leopoldo, Unisinos, 2011.

Nunes, C., "Cooperativas, uma possível transformação identitária para os trabalhadores do setor informal," *Sociedade e Estado*, 2001, Vol. 16, No. 1-2, pp. 134-158.

Pamplona, J. B., "A controvérsia conceitual acerca do setor informal e sua natureza político-ideológica," *Cadernos PUC*, São Paulo, 2001, No. 11, pp. 11-78.

Petitclerc, M., *Nous protégeons l'infortune. Les origines populaires de l'économie sociale au Québec*, Montreal, VLB Éditeur, 2007.

Pinto, J., *Economia solidária; de volta à arte da associação*, Porto Alegre, UFRGS, 2006.

Razeto, L., *Las empresas alternativas*, Montevideo, Editorial Nordan-Comunidad, 1990.

Singlemann, P., "Political Structure and Social Banditry in Northeast Brazil," *Journal of Latin American Studies*, 1975, Vol. 7, No. 1, pp. 59-83.

Sousa Santos, B., "Porque é tão difícil construir uma teoria crítica?," *Revista Crítica de Ciências Sociais*, 1999, Vol. 54, pp. 197-215.

——, "A Critique of Lazy Reason: Against the Waste of Experience," in Wallerstein, I. (ed.), *The Modern World-System in the Longue Durée*, London, Paradigm Publishers, 2004, pp. 157-197.

Sousa Santos, B. and Meneses, M. P., *Epistemologias do Sul*, Coimbra, Almedina, 2009.

Souza, J., *Possibilidades e limites da associação na estruturação de unidades locais de reciclagem. O caso da Associação NORA – Novo Osasco Reciclando Atitudes – dos trabalhadores com materiais recicláveis*, Master's dissertation in Applied Social Sciences, Universidade do Vale do Rio dos Sinos, 2005.

Wautiez, F., Soares, C. and Lisboa, A., "Indicadores de economia solidária," in Cattani, A. (ed.), *A outra economia*, Porto Alegre, Veraz, 2003, pp. 281-291.

To Estimate the Scope and Size of the Social Economy in Japan

Challenges for Producing Comprehensive Statistics

Akira KURIMOTO

Professor, Institute for Solidarity-based Society
at Hosei University, Tokyo, and director
of the Consumer Co-operative Institute of Japan

Introduction

The social economy in Japan has made substantial growth and played significant roles in some socio-economic branches. The nonprofit sector has been the principal provider of social and health services, and education at various levels. The co-operative sector has played a pivotal role in handling the bulk of foodstuff. Agricultural co-operatives are ranked as the largest co-operative organizations in the ICA's Global 300 while consumer co-ops have grown to be powerful consumer organizations with 26 million members comparable to the European counterparts.

However, the social economy is not visible in comparison with the powerful corporate sector and the commanding public sector. It lacks the identity as a sector and the cohesion among organizations involved that resulted in low recognition by the government, media or academism. There are several reasons to explain this lack of visibility but the most important one is the institutional divide. The lack of the comprehensive statistics has been a serious flaw in the public recognition. The Democratic Party of Japan (DPJ) government initiated a round table on "New Public Commons" which proposed to promote the "Civic Sector" that largely corresponded to the social economy for the first time in the official documents but there is no follow up by the returned Liberal Democratic Party (LDP) government.

The government-sponsored Research Institute of Economy, Trade & Industry (RIETI) conducted comprehensive surveys to grasp the whole picture of the Third Sector in 2010 and 2012. They covered a wide range of broadly defined non-profit sector as well as co-operatives and neighborhood associations. Through these surveys we could learn about the characteristics and challenges of the social economy organizations (SEOs)

This paper will illustrate a sheer size of the social economy in Japan and explain why its sector is not visible from legal-administrative and political economy contexts. It will give a short description on the scope and size of SEOs based on the existing statistics. Then it will show the methodology, main findings and policy implications of RIETI study on the Third Sector. Finally, it will propose some recommendations for producing comprehensive statistics of SEOs.

1. Sizable social economy in Japan

The social economy in Japan has made substantial growth in the state-led capitalism and played significant roles in some socio-economic branches. In the nonprofit sector, social welfare corporations have been assigned the role of principal provider of social services for elderly, child and the handicapped while medical corporations occupy predominant position in providing health services. Private school corporations cover more than 70% of higher education. The Kobe earthquake in 1995 sparked voluntarism among a great number of citizens who rushed to help victims. Such surge of volunteers resulted in the enactment of the Act to Promote Specified Nonprofit Activities (so-called NPO Act) in 1998. Accordingly newly born specified NPOs mushroomed in the last decade and surpassed 40,000 organizations. These new and older nonprofit organizations account for ca. 5.2% of GDP and 10.0% of employment in the Satellite Account of Non-Profit Institutions (NPI) (Salamon *et al.*, 2013). It means their contribution to GDP is smaller than Canada (8.1%) and the USA (6.6%), but larger than Australia (4.9%) and France (4.7%).

In the co-operative sector, agricultural co-operatives have played a pivotal role in handling the bulk of foodstuff (63% of beef, 50% of rice etc.) and exercised a strong influence on the agricultural policy. They are among the largest co-operative organizations in the world as shown in the ICA's Global 300 ranking; Zen-Noh (supply & marketing) as No. 1, Zenkyoren (insurance) as No. 2 and Norinchukin Bank as No. 12 in 2006. They are among the largest trading companies, insurance providers and banks in Japan. Consumer co-ops have grown to be powerful consumer organizations with 26 million members (embracing more than 40% of households) and the turnover of JPY 3.3 trillion (accounting for 5.8% of food retailing), which means they are collectively the third largest retailer after Aeon and Ito-Yokado groups. Their membership and turnover accounted for 97% and 37% of EUROCOOP member co-ops in 2010, which means the Japanese consumer co-ops have grown much bigger than the European counterpart. Insurance co-ops occupy 23% of life insurance contracts while co-operative banks handle 24% of savings and 18% of loans. Thus co-operatives present the enormous size both in domestic and international terms.

2. Institutional divide of social economy

However, the social economy is not visible in comparison with the powerful corporate sector and the commanding public sector. It lacks the collective identity as a sector and the cohesion among organizations involved, which resulted in low recognition by the government, media or academism. There are several reasons to explain this lack of visibility but the most important one is the institutional divide.

The traditional nonprofits for health, welfare and education have been controlled in the varied degree by different ministries. For instance, social welfare corporations, that are given tax privileges and public money for construction and operation, are often seen as quasi-autonomous non-governmental organizations (QUANGOs) accepting retired bureaucrats as the executives while medical corporations are subject to strict regulations for medical care but not given tax privileges. Newly born grass root specified NPOs had no tax incentives and remained very small both in size and impact. The annual budget of more than a half of them has been less than JPY 5 million which means they can't employ full-time staff. So the nonprofit sector is bipolarized between traditional nonprofits with stronger financial resources linked with higher dependence to governments and newly-born nonprofits with opposite traits. The interest groups such as medical associations are actively promoting specific interest of individual industries while there is no umbrella organization representing the whole nonprofit sector such as NCVO in the UK.

Co-operatives have been divided by separate laws and supervising ministries.[1] Agricultural co-ops had been promoted as agents for implementing agricultural policy and given a wide range of support by the government, while they had been protected from competition. They were seen as combination of government's subcontractors, pressure groups and co-ops per se. In contrast, consumer co-ops had been handicapped by strict restrictions including complete prohibition of non-member trade and inter-provincial trade, which were introduced under the pressure of small retailers. Thus, the protectionist policies had

---

[1] The uniform Industrial Co-operative Act of 1900 was replaced by dozens of co-operative laws according to needs of the industrial policies since 1945.

affected agricultural co-ops and consumer co-ops in opposite vectors and contributed to distinctive organizational culture and political affiliation. The workers co-ops lack legal recognition despite campaigning for decades. Co-operatives are also compartmentalized based on industrial policies and there is neither comprehensive statistics that transcend various types of co-ops nor coordinating body that represents the entire co-operative sector.

Such divide can be explained in legal-administrative system and political economy. In the Japanese legal system, Article 33 of the Civil Code of 1896 required that all legal persons be formed in accordance with its regulations (*Hojin hoteishugi*). Article 35 provided for the establishment of for-profit organizations, while Article 34 provided for only the Public Interest Corporations (PIC), relating to academic activities, art, charity, worship, religion, or other public interest.[2] As Pekkanen (2000) argues, "This creates a legal blind spot – most groups that are nonprofit but not in the 'public interest' had no legal basis whatever to form". A number of laws were created for nonprofit organizations catering to the specific social needs while the specified NPO Act was enacted in 1998 to fill the gap pointed out by Pekkanen.

The corporate forms correspond to the supervising authorities which mandates are provided by specific laws (See Table 1). The SEOs are classified under the category of "not-for profit". In contrast to for-profit companies that are to be incorporated by only registering at the District Legal Affairs Bureaus and allowed substantial autonomy in the business administration, the not-for profit and public corporations are subject to the strict supervision for a wide range of matters including incorporation, governance and financing. They are incorporated by the jurisdictional authorities' conducts such as authorization, approval, licensing or certification while their governance and financial matters are precisely stipulated in a set of laws, ordinances, instructions or administrative guidance. They are so dependent to the public policy and administrative discretion that they have developed strong pressure groups to protect their specific interests. This is why the strongest pressure groups such as agricultural co-ops and medical associations are found among these not-for profit organizations.

---

[2]   Currently these Articles are integrated in Article 33.

Table 1. *Types of corporations in Japan*

Category	Corporate bodies	Legislation	Supervisory authority
For profit	Joint-stock Company	Companies Act	Corresponding Min. for functional matters
	General Partnership Company		
	Limited Partnership Company		
	Limited Liability Company		
Not-for profit	Public Interest Corporation	Act on Authorization of PICs	Cabinet Office
	General Corporation	Act on General Corporations	Cabinet Office
	Social Welfare Corp.	Social Welfare Act	MHLW
	Medical Corp.	Medical Service Act	MHLW
	Private School Corp.	Private School Act	MEXT
	Religious Corp.	Religious Corp. Act	MEXT
	Specified Nonprofit Corp. (NPO)	Act to Promote Specified Nonprofit Activities	Cabinet Office
	Agricultural Co-op	Agricultural Co-op Act	MAFF
	Fishery Co-op	Fishery Co-op Act	MAFF
	Forestry Co-op	Forestry Co-op Act	MAFF
	Consumer Co-op	Consumer Co-op Act	MHLW
	SME Co-op	SME Co-op Act	METI, MOF
	Shinyo Kinko (Sinkin)	Sinkin Act	MOF
	Labor Bank	Labor Bank Act	MHLW, MOF
	Trade Union	Trade Union Act	MHLW
Public	Local Government	Local Autonomy Act	MIC
	Independent Administrative Corp.	General Rule Act for IAC	Corresponding Min.
	National University Corp.	Act of National University Corporations	MEXT

Corp: Corporation, Co-op: Co-operative, MHLW: Ministry of Health, Labor and Welfare, MOF: Ministry of Finance
MAFF: Ministry of Agriculture, Forestry and Fisheries, MEXT: Ministry of Education, Culture, Sports, Science and Technology
METI: Ministry of Economy, Trade and Industry, MIC: Ministry of Internal Affairs and Communications

The institutional reform on the Public Interest Corporation (PIC) has been made in the new millenium. The existing PIC system had been criticized because the competent supervisory authorities (ministries and prefectures) that had a wide range of discretional power in incorporation and ongoing supervision often used PIC as a vehicle to pursue their own interests by channeling public subsidies and securing high-ranking posts for *amakudari* (practice of parachuting retired government officials onto public corporations or private businesses). Upon groundwork for reforming corporation and taxation system, three bills had gotten through the Diet in 2006 and put into force in 2008. They are the Act on General Incorporated Associations and Foundations, the Act on Authorization of Public Interest Incorporated Associations and Foundations and the Act on Adjustment of Relating Acts pertaining to the Enforcement of affore-mentioned Acts that absorbed the Intermediary Corporation Act of 2001 for incorporating nonprofit and non-public interest organizations such as sports clubs and alumni associations. Under the new law, the incorporation and charitable status was separated. People can set up a general incorporated association or foundation simply by registration at the District Legal Affairs Bureaus without any kinds of governments' permission while the newly created independent body called Public Interest Corporation Commission (PICC) shall give charitable status by authorizing the public interest corporations (association or foundation) which meet requirements provided in the law. Originally it was sought to enact the general nonprofit organization law including specified NPOs but non-dintribution constraint was not stipulated and the specified NPOs Act continued to exist.

Since the government has the authority in making and implementing the industrial policy, the trade associations (*gyokai dantai*) on the top of hierarchy in each industry try to pursue their specific interests by approaching the appropriate bureaus of relevant ministries or mobilizing the power of politicians. So, the coalition structure emerged among three parties so long as they share common interests. The trade associations often accept state subsidies and ex-bureaucrats to remunerate to the special benefits they can enjoy and lobby the MPs to promote specific interests of particular industries through contributing both votes and money. Such tripartite relationship of the MPs of LDP (*zoku giin*), the business sector and the bureaucracy in post-World War II Japan is often described as "iron triangle" as the special trait of the Japanese political economy. Aoki puts it as "compartmentalized pluralism" or "bureau pluralism" (Aoki, 2011). In the social economy, agricultural co-ops and medical associations have developed such coalition with ministries and LDP in shaping industry-specific public policies to promote their interests.

In addition, the notion of the Social Economy is not agreed among researchers and there is a competition of different brands; voluntary sector, non-profit and co-operative sector, civic sector, social sector and so on. Also, the lack of the comprehensive statistics has been a serious flaw in obtaining the wider recognition. The social economy enterprises were often referred as Community Business or Social Business, both of which were used by METI that intended to enclose them within its territory but had no political initiative to institutionalize them. There is still no public policy addressed to the social economy. The DPJ government initiated a round table on "New Public Commons" in 2009 which proposed to promote "the Civic Sector" for the first time in the official document that largely corresponds to the social economy.[3] It implemented some measures to encourage community businesses with special emphasis on the reconstruction of areas affected by the earthquake and tsunami during 2011-2012[4] but there is no follow-up action after the return of the LDP-led government in 2012.

The research on the Third Sector has been divided between American nonprofit organization school and European social economy school. The former is focusing on nonprofits, volunteering and social entrepreneur while the latter is disseminating notions of social economy and social economy enterprises including co-operatives and mutual organizations. The communication between these schools is very limited while individual researchers are often pursuing their own agenda paying little attention to the others. The Japan NPO Research Association (JANPORA) has very limited focuses on the specified NPOs while the Japanese Society for Co-operative Studies (JSCS) is dominated by agronomists and agricultural co-op officers.

3. Scope and size of social economy[5]

There are several problems in compiling statistics on social economy organizations (SEOs). First, we have to decide on the scope of entities that can be classified as SEOs. There is no doubt that nonprofits and co-operatives regulated under the different laws are included, but there is a unique problem pertaining to the qualification of mutual. The mutual companies are authorized by the Premier's licenses under the Insurance Business Act of 1946. Originally co-operatives wished to provide insurance for their member's life and properties even before the Second World War. But they were not allowed to do so because of the strong resistance from insurance companies that intended to protect their vested interests. So they had to organized co-operative insurance or "kyosai" that were

---

[3] The Civic Sector consists of specified NPOs, incorporated association/foundations, private school corp., social welfare corp., medical corp., co-operatives, unincorporated neighborhood associations and for-profit organizations mainly working for public interests.

[4] The project to support "New Public Commons" (ca.JPY 10 billion) and the project to create community employment (ca.JPY 3 billion) mainly targeted specified NPOs.

[5] This section was originally presented in Kurimoto, 2011.

regulated under the different co-operative laws and ministries. Today five life insurance companies are operating as mutual companies[6] while others were demutualized since 2002. They have not seen themselves as a part of the social economy and had often confronted with consumer criticism because of their behavior prioritizing profits over policy holder's benefits. Although the ICMIF included these companies in its statistics, it indicated a legend "regardless of historical circumstances of establishment" according to the request of the Japan Co-operative Insurance Association.

The widest definition may include unincorporated entities such as neighborhood associations (urban *chonaikai, jichikai* or rural *shuraku*). They cover most of households by automatic membership, which has been often seen as contrary to voluntary membership principle of SEOs. They exist at every corner of the country but most of them are not incorporated.[7] They are generally seen as the administrative tail ends and gradually fading out in both urban and rural areas. The trade unions have been organized as enterprise unions and once had a strong influence in raising wages during the rapid economic expansion period since mid-fifties. But they could not stop the continuous decline in membership (now covering only 18% of workforce) nor resist businesses' survival strategy that generated the massive precarious workers.

The comprehensive statistics of SEOs is not yet available due to various reasons. The Economic Census compiled by the government's Statistics Bureau provides the most complete enumeration of businesses on their structure and economic activities. The for-profit corporations are classified in detail according to the corporate forms (joint stock company, general and limited partnership company, LLC, mutual company etc.), while nonprofits, co-operatives and others are crowded into a single category "the other legal persons,"[8] which largely hamper the scientific analysis.[9]

The size of nonprofit institutions (NPI) was calculated by the Economic and Social Research Institute according to the *UN Handbook*

---

[6]   The World Co-operative Monitor in 2012 enlists Nippon Life (5th), Meiji Yasuda Life (10th), Sumitomo Life (11th), Fukoku Life and Asahi Life in the Global 300.

[7]   Their origin can be dated back to *gonin-gumi* (group of five households) during the Tokugawa period but in the modern history they were organized for mutual help and/ or surveillance among neighbors during the Second World War. They were banned by the GHQ and reorganized as autonomous associations in 1952.

[8]   The questionnaire exemplified public corporations, incorporated associations/ foundations, private school corp., social welfare corp., religious corp., medical corp., trade unions, agricultural/fishery co-ops, business co-ops, mutual societies, national health insurance societies, sinkin banks etc. as "the other legal persons".

[9]   Hence the utility of the satellite account methodology. See in the book the chapter signed by Fecher and Ben Sedrine-Lejeune.

in 2004. It included social welfare corporations, medical corporations, private school corporations, religious corporations, public interest corporations, specified NPOs and unincorporated entities. Core monetary valuables such as market output, intermediate consumption and added values as well as structural valuables such as paid employment and membership were calculated based on the existing data sources. To grasp the characteristics of NPI, such valuables as non-market output and volunteer employment were also estimated. Then tables of the NPI satellite account were created for the aggregate NPI sector and groups based on the International Classification of Non-Profit Organizations (ICNPO).

Table 2. *Organizations included in the NPI satellite account and data sources*

Scope	Corporate form	Data sources to tap information
Narrow	specified NPOs, unincorporated entities	Basic Survey on Civic Organizations
Wider	+ public interest corporations, social welfare corporations, medical corporations, private school corporations, religious corporations etc.	Survey on PICs Survey on Private Nonprofits Today's Finance of Private Schools Survey on Medical Economy

These nonprofit organizations' GDP (JPY 20.7 trillion) accounts for approx. 4.2% of the Japanese GDP, which meant they were much larger than electromechanical industry (JPY 16.7 trillion) and communication services (JPY 10.8 trillion). They had largely increased from 2.2% in 1990. Amongst added values, health services accounted for more than a half (55.4%) to be followed by social services (18.4%) and education/ research (17.8%). Their contribution to the FTE based employment was 7.3% (employees) and 3.6% (volunteers).

Table 3. *Japanese non-profit institution's share in GDP in 2004*

	Added value (JPY trillion)	Share in nominal GDP
NPI based on SNA	20.7	4.2%
NPI incl. Non market output	24.1	4.8%
NPI incl. Non-market output & volunteers contribution	28.7	5.8%

Source: Economic and Social Research Institute, Cabinet Office, *National Accounts Quarterly*, No. 135, 2008.

The diverse and fragmented statistics of co-operatives made it difficult to grasp their overall picture and gravity of the sector in the national economy. The MAFF had compiled extremely detailed statistics on agricultural, fishery and forestry co-ops as a part of the industrial statistics on the primary industries that had been disproportionately large. On the other hand, the MHLW had very limited official statistics on consumer co-ops despite of their sheer size. Neither the ministry nor the trade associations had official statistics on the small and medium sized enterprises co-ops while emerging worker co-ops are not yet institutionally recognized and therefore had no official statistics. The aggregated data of official statistics are shown in Table 4 but the total membership of nearly 100 million is nearly equivalent to the whole adult population since it inevitably implies substantial overlapping among the same and different kinds of co-ops. For instance, having membership of several co-ops is quite common practice in consumer co-ops, of which official statistics have not taken into account. It was not envisaged to count the number of households affiliated with co-ops. Some national co-op federations have compiled much detailed statistics for managerial purposes but they have also some problems associated with different methodologies and inclusiveness.

Table 4. *Key figures of Japanese co-operatives*

	No. of co-ops	Membership (0000)	Employees (0000)	Turnover (JPY bil.)	Year
Agriculture	844	943	23	7,626	2007
Fishery	2,747	40	1	1,373	2007
Forestry	736	159	1	31	2007
Consumer	1,093	6,318	12	3,541	2007
SME	na	na	na	na	
Other credit	292	1,914	13	na	2008
Other insurance	4,105	442	0	na	2008
Workers	na	na	na	na	
Total	9,817	9,816	49	12,571	

Source: Kurimoto (2011) out of official statistics.

4. RIETI surveys on the Third Sector

The government-sponsored RIETI (Research Institute of Economy, Trade & Industry) set up a Study Group (chaired by Prof. Fusao Ushiro, Nagoya University) aiming to grasp the whole picture of the Third Sector in 2010. Compared to previous discussions on non-profit organizations that focused only on specified NPOs, this third sector corresponds to the social economy and comprises the broadly defined non-profit sector as well as co-operatives, neighborhood associations, and social economy enterprises.

4.1. Methodology of surveys

We conducted the demographic surveys on the characteristics and management problems of the Third Sector in 2010 and 2012.[10] We targeted organizations classified in "the other legal persons" and "unincorporated entities" in the Economic Census.[11] In 2009 these organizations belonging to broadly defined third sector accounted for 6.9% of businesses and 11.5% of employees, steadily expanding from 5.4% and 9.1% in 1999. In 2010 the questionnaires were sent to 12,500 organizations out of which valid responses were obtained from 3,901 (31.2%). In 2012, 3,656 organizations out of 14,000 (26.1%) made valid responses. These organizations were selected by stratified random sampling; first, the targeted number of samples were distributed pro rata to the population of corporate forms, then the samples were chosen at random from each groups. The questionnaires were resent to the valid addresses of organizations for bouncing mails.

Table 5. *Population and samples for 2012 Survey*

	Population*	Samples
The other legal persons	163,109	11,900
Unincorporated entities	29,245	2,100
Total	192,354	14,000

*Based on 2009 Economic Census.

The responded organizations in the second survey were as shown in table 6. The share of respondents of general and public interest associations/foundations and specified NPOs increased while that of agricultural and consumer co-ops decreased to the extent that questioned representativeness of obtained data but no corrective measures were taken because of constraints in time and costs. No matching test was conducted either.

---

[10]   RIETI outsourced dispatching questionnaires and tallying responses to Tokyo Shoko Research, Ltd.

[11]   The Economic Census is a survey to obtain fundamental statistics of economic structure stipulated by the Statistics Act (No. 53, 23 May 2007). It is sought to establish comprehensive statistics by integrating existing business and industrial statistics. Data is collected through investigator's visits and mailing designated by all municipalities commissioned by prefectures and central government (MIC and METI). All businesses are obliged to respond by the Act. The first survey in 2009 collected data from ca. 6.36 million businesses.

Table 6. *Number and share of responding entities 2012*

	No. of respondents	Share (%)
General incorporated associations	127	3.5
General incorporated foundations	60	1.6
Public interest incorporated associations	113	3.1
Public interest incorporated foundations	141	3.9
Associations/foundations in transition	325	8.9
Social welfare corporations	513	14.0
Private school corporations	126	3.4
Medical corporations	87	2.4
Specified nonprofit corporations (NPOs)	378	10.3
Vocational training corporations	125	3.4
Offenders rehabilitation corporations	68	1.9
Consumer co-operatives	26	0.7
Agricultural co-operatives	41	1.1
Fishery co-operatives	61	1.7
Forest owners co-operatives	41	1.1
SME co-operatives	328	9.0
Financial co-operatives	155	4.2
Insurance co-operatives	25	0.7
Public corporations	88	2.4
Other corporations	422	11.5
All corporations	3,250	88.9
Unincorporated	406	11.1
Grand total	3,656	100.0

The questionnaires were designed by the RIETI Study Group aiming to grasp the management problems of the organizations at micro levels. The basic structure of questionnaires was not changed to facilitate the effective comparison while the second survey increased types of entities to grasp more detailed picture of the sector. They covered the following questions;

I. On the organizations

Corporate forms; number of business places; number and remunerations of officers; age, remunerations and career of CEOs; number and profession of auditors; number and remunerations of employees; number and hours of volunteers; conditions of employment; number and method of recruitment; training system; year of establishment and incorporation; support on the establishment.

II. On the governance

Disclosure of information; frequency of meetings; mode of audit (internal or external).

III. On the activities

Field of activities; nature of activities; area of operations.

IV. On the finance

Composition of expenses (transfer, personnel and indirect costs); composition of income (earned and voluntary); income through voucher system and related business.

V. On future perspectives

Intension of expansion in field of activities, area of operations, employees and finance.

The responses to the questionnaire were tallied to create tables that illustrated the characteristics of the third-sector organizations according to corporate forms including cross analyses that presented the correlation between fields and modes of activities, e.g. personal care with business operation, academic promotion with research institutions, international exchange with granting etc.

4.2. Main findings of surveys

Since there was no significant changes in the results of surveys, herewith the main findings of the second survey are shown as follows.

- 84.5% of respondents had paid employees while 30% of specified NPOs had no paid employee. The average number of employees was 41.6 while the median was only 4.0. The median of highest annual income was JPY 8.77 million while that of lowest income was JPY 1.35 million. 65.1% of respondents had part-time employees.

- 16.5% of respondents had unpaid volunteers while the median number was 11 and the median working hours were 6 hours per month. 6.3% of respondents had paid volunteers while the median number was 7 and the median working hours were 15 hours per month with a pay of JPY 800 per hour.

- Fields of activities depended on the purpose of organizations and therefore were extremely diversified; economic activities (15.9%), child care (9.5%), elderly care (9.0%) support to the handicapped (8.5%), local development (7.8%), vocational training (4.6%), health care (4.1%), supply of safe foods (3.7%) and long tail follows.

- 60 to 75% of respondents disclosed information (bylaws, annual reports and financial statements) to the concerned and at offices while only 18 to 27% disclosed information on Internet. Financial co-ops and public interest foundations are among the highest use of Internet.

- 97.7% of respondents conducted auditing. 87% of them had internal audits while 30% had external audits by specialists such as certified and/or tax accountants. Generally speaking, the larger organizations conducted external audits.

- The median expenditure was JPY 50.36 million ranging from JPY 6.64 million (unincorporated) to JPY 1,969.87 million (financial co-ops). The median of specified NPOs was only JPY 15.67 million, that means they could have very few fulltime employees. The expenditure consists of transfer (63.6%), personnel costs (13.4%), other costs (15.0%) and indirect costs (8.0%).

The other interesting finding was the varied income structures in terms of sources and nature. The sources of income consist of individual citizen, public sector, for-profit corporate sector and the third sector. The nature of incomes is either earned income or voluntary income. The former includes sales of goods and services, fee from commissioned business or remuneration of services from voucher system[12] while the latter includes grants, donation or subscription. The contribution of the public sector is highest in medical corporations (95.7%). This doesn't mean they were heavily subsidized by the public money but paid through the compulsory medical insurance system. In case of social welfare corporations, the public money accounted for 88.6%, out of which 78.8% came from long-term care insurance while 9.8% was given as subsidies. As far as the nature of income in the nonprofit sector is concerned, the earned income accounted for 69%, much larger than the voluntary income. It was much larger in comparison with the United States and the UK, indicating that shifting "from grants to contracts" has been already underway to a large extent. On the contrary, individual and corporate donations were far smaller in comparison with those countries. In the consumer co-operatives, the earned income through trading activities in the market accounted for 99%.

Table 7. *Source and nature of income of the third sector (%)*

Source of income	Earned	Voluntary	Others	Total
Individual citizens	16.3	8.6		24.9
Public sector	23.2	7.6		30.8
Corporate sector	15	0.6		15.6
Third sector	6	2.5		8.5
Others			19.8	19.8
Total	60.5	19.3	19.8	99.6

---

[12] Here the voucher system is used in a wider sense in that service users can choose service providers who are redeemed service expenses from the public insurance scheme such as medical and long-term care insurance.

Table 8. *Source and nature of income of the nonprofit sector (%)*

Source of income	Earned	Voluntary	Others	Total
Individual citizens	11.7	3.9		15.6
Public sector	50.5	10.5		61.0
Corporate sector	5.3	1.6		6.9
Third sector	1.7	1.1		2.8
Others			13.8	13.8
Total	69.2	17.1	13.8	100.1

We can locate the third sector organizations by the extent of public money and nature of income. Here "higher" or "lower" means more or less than average. Dimension I includes organizations which earn income from public money through vouchers and commissioned businesses. Medical corporations, social welfare corporations and specified NPOs fall in this category. Dimension II includes rather few organizations which receive public money through grants and donations. Dimension III includes organizations which earn income from private sector. Co-operatives are typical organizations trading in the market while general associations/ foundations and public interest associations/foundations falls into this category to a lesser extent. Dimension IV includes organizations which earn income from private sector while receiving substantial donations from individuals (parents, graduates etc.). Private school corporations fall into this category. Thus the income structure clearly demonstrated the characteristics of the third sector organizations.

Figure 1. *Positioning of the third sector organizations by source of income*

	←Higher public money	Lower public money→
↑ Higher earned income	I Medical corp. Social welfare corp. Specified NPOs	III Co-operatives General foundations General associations Public interest association Public interest foundation
Lower earned income ↓	II Vocational training corp. Offender rehabilitation corp. Other corp.	IV Private school corp. Unincorporated

4.3. Some policy implications

This survey revealed a vast variety of the third sector organizations in terms of scope and size of activities, which was useful to grasp the overall pictures on the organizational capacity, inter alia their income structure.

It was the first attempt to collect transversal information on the sector as a whole. In addition, this approach highlighted the characteristics of the organizations according to corporate forms and provided information that can be used to design the institutional reforms aiming at enhancing their capacity. To collect information from the organizations based on questionnaires required to have substantial number of samples. The number of samples of co-operatives must be increased since it was too small to collect the reliable data sets in comparison with nonprofits.

Although this survey has not yet given tangible impacts to the public policy and sector mobilization in Japan, we can draw some policy implications. First of all, we observe the rapid increase of the general corporations (associations and foundations). The responses increased from 84 to 187 in 2010-2012. They numbered ca. 24,000 as of August 2012 with increase of ca. 9,500 during a year. As mentioned earlier, the PIC reform in 2006 was intended to create the standard nonprofit organization law covering former PICs, specified NPOs and intermediary corporations. However, the priority was placed to reduce the government-sponsored corporations created by competent agencies, staffed with seconded civil servants and financially supported with public money. Finally it ended up as a half-baked reform in that non-distribution constraints were not stipulated as default and non-fiscal incentives were introduced except for the authorized corporations. Since then the general corporations have been established at a rapid pace and may outnumber the specified NPOs in a few years. It means people choose them that can be set up without government's conducts rather than specific NPOs that need to obtain official certification. The former are more varied in size while the latter remained very small since the inception in 1998 despite a large amount of public money spent for promoting them.[13] The general corporations seem to become the dominant form of nonprofit organization in the near future if the appropriate support structure is introduced. In addition, the PIC reform should be followed up by reforms of other types of nonprofit organizations including social welfare corporations, private school corporations and medical corporations. It is also necessary to design the co-operative framework law that will facilitate the incorporation of general co-operatives free from ministries' control. By doing so, we can change the old structure characterized by the institutional divide and create the visible social economy.

---

[13] The specified NPOs have been given a variety of supports by governments at various levels. Most of ca. 300 NPO centers have been given subsidies or commissioned for services.

At the same time, the institutional divide can be reduced by lowering the barrier to entry to some social services. For instance, the elderly nursing homes can be built and operated only by social welfare corporations that fulfill the requirements stipulated by regulations although non-residential elderly care services were opened up to various kind of providers including for-profit and non-profit entities under the Long Term Care Insurance Act effectuated in 2000. Such barrier had been introduced to facilitate the government's control of subordinate organizations and largely hampered the effective competition and innovation among service providers.

It was also demonstrated that the older nonprofits and co-operatives have much larger resources (finance, manpower, experiences and so on) to be used to establish new organizations to meet emerging socioeconomic needs of people and contribute to the enhancement of the social economy as a whole. They can be rejuvenated by helping to create start-ups, spinning off the social economy enterprises and supporting to create networks among SEOs. For instance, co-operatives have made outreach to supply food to people in under populated area or disaster-stricken area. They also helped to set up specified NPOs and social economy enterprises to cope with social exclusion of the handicapped, heavily-indebted and homeless people, ethnic minorities and so on. Such collaboration between "older" and "newer" SEOs can contribute to the advancement and visibility of the social economy.

Conclusion: challenges for producing comprehensive statistics

As mentioned earlier, the social economy is divided by the legal-administrative system and political economy. We are facing a couple of challenges for producing comprehensive statistics of the Japanese SEOs. So, the following recommendations are to be considered.

First of all, we have to discuss on the qualifications of SEOs and methodologies to create cross-cutting statistics that can be widely used by policy makers and researchers. RIETI surveys transcending nonprofits and co-operatives can be a starting point to understand the general trends of SEOs according to the corporate forms. It should be ameliorated by improved methodologies and representativeness of each corporate form.

Secondly we have to create a Satellite Account of Co-operatives to grasp their weight in the national economy so that we can have a Satellite Account of SEOs. We have already had the Satellite Account of Non Profit Institutions calculated in accordance with the *UN Handbook*. CIRIEC published a *Manual for drawing up the satellite account in the Social Economy: Co-operatives and Mutual Societies* in 2006. This initiative should be followed up by the UN agencies such as ILO and FAO. In this regard, we have to discuss on the treatment of mutuals. The ICMIF's Global 500 reports include the Japanese life insurance companies that have the corporate status of 'mutual companies' based on the Insurance Business Act. But the co-ops argue that these companies have had no contact with co-ops and often confronted with the latter; they opposed co-ops to enter the insurance business so co-ops had to establish insurance business under respective co-operative laws. Consumer movement has criticized that insurance companies had often neglected policyholder's rights, prioritizing the interests of their executives over consumers. So the Japan Co-op Insurance Association (trade association of insurance co-ops) asked the ICMIF to indicate the legend that the Global 500 did not take the historical circumstances of establishment into consideration. Although these companies have very little common identity or contact with co-operatives, we have to solve this question to create internationally comparable statistics of the social economy.

Thirdly, we have to improve the Economic Census that was commenced in 2009 to grasp the comprehensive picture of the economic activities, integrating all the economic statistics that had been prepared by different ministries in line with the industrial policies (commerce, manufacturing, mining, service industry, etc.). It covers ca. 6.36 million businesses and will be used to the calculation of GDP, or formulation of local development policies, and so on. Although nonprofits, co-operatives and others are included in a single category "the other legal persons," such distinguished corporate forms should be classified separately to produce reliable statistics of the social economy. The RIETI Study Group plans to submit a request to the Statistics Bureau that is responsible for implementing the Census to add a question on the concrete corporate forms or make individual data available to create sectoral statistics.

References

Aoki, M., "Beyond Bureau Pluralism," *Opinions*, Japanese Institute of Global Communication, Global communications platform, 2011, http://www.glocom. org/opinions/essays/200109_aoki_beyond/ [accessed June 8, 2015].

Barea, J. and Monzón, J. L., *Manual for Drawing up the Satellite Account in the Social Economy: Co-operatives and Mutual Societies*, CIRIEC, 2006.

Borzaga, C. and Defourny, J. (eds.), *The Emergence of Social Economy Enterprise*, London, Routledge, 2001.

Chavez, R. and Monzón, J. L., *The Social Economy in the European Union*, CIRIEC, 2007.

Deller, S. *et al.*, *Research on the Economic Impact of Co-operatives*, Madison, WI, University of Wisconsin Center for Cooperatives, 2009.

Economic and Social Research Institute, Cabinet Office, *National Accounts Quarterly*, No. 135, 2008.

Kurimoto, A., "Evaluation of Co-operative Performances and Specificities in Japan," in Bouchard, M. J. (ed.), *The Worth of the Social Economy: An International Perspective*, Brussels, Peter Lang, 2009, pp. 213-244.

———, *Divided Third Sector in the Emerging Civil Society: Can DPJ Contribute to Changes?*, paper presented for workshop "Continuity and Discontinuity in Socio-Economic Policies in Japan: From the LDP to the DPJ," University of Sheffield, 4 March 2011.

———, "Towards Statistical Grasp of the Social Economy in Japan," in Osawa, M. (ed.), *Future Developed by the Social Economy*, Minerva Shobo (Japanese), 2011.

Le Grand, J., *The Other Invisible Hand. Public Services through Choice and Competition*, Princeton University Press, 2007.

OECD (ed.), *The Changing Boundaries of Social Economy Enterprises*, OECD, 2009.

Pekkanen, R., "Japan's New Politics: The Case of the NPO Act," *Journal of Japanese Studies*, Vol. 26, No. 1, 2000, pp. 111-148.

Pestoff, V. A., *Beyond the Market and State: Social Economy Enterprises and Civil Democracy in a Welfare Society*, Alderhot, Ashgate, 1998, 2005.

Salamon, L. M. and Anheier, H. K., *The Emerging Sector: An Overview*, The Johns Hopkins University Institute for Policy Studies, 1994.

Salamon, L. M., Haddock, M. A., Sokolowski, S. W. and Tice, H. S., *Measuring Civil Society and Volunteering: Initial Findings from Implementation of the UN Handbook on Nonprofit Institutions*, Baltimore, Johns Hopkins University, Center for Civil Society Studies, Working Paper No. 23, 2007.

Salamon, L. M., Sokolowski, S. W., Haddock, M. A. and Tice, H. S., *The State of Global Civil Society and Volunteering, Latest Findings from the Implementation of the UN Nonprofit Handbook*, Baltimore, Johns Hopkins University, Center for Civil Society Studies, Comparative Nonprofit Sector Working Paper No. 49, March 2013.

United Nations, *Handbook on Non-Profit Institutions in the System of National Accounts*, New York, United Nations, 2003.

Ushiro, F., "Scope and Current Situation of the Third Sector in Japan," *RIETI Discussion Paper Series*, RIETI (Japanese), 2011.

———, "Current Situation and Challenges of the Third Sector Organizations in Japan," *RIETI Discussion Paper Series*, RIETI (Japanese), 2012.

———, "Current Situation of the Third Sector Organizations and Challenges to Build the Sector," *RIETI Discussion Paper Series*, RIETI (Japanese), 2013.

Challenges in Conducting a Study on the Economic Impact of Co-operatives

Nicoleta UZEA

Postdoctoral research associate, Ivey Business School,
Western University, Canada

Fiona DUGUID

Member of the board, Canadian Association for Studies
in Co-operation, Community co-lead, National Study
on the Impact of Co-operatives

1. Introduction

The resilience of cooperatives in the aftermath of the 2008 global economic crisis[1] (Birchall and Ketilson, 2009; Roelants *et al.*, 2012; Birchall, 2013) has led to increased interest in the cooperative business model, and the impact that cooperatives have on the economy and the communities in which they operate. In 2009, the University of Wisconsin Center for Cooperatives produced the first comprehensive set of national-level statistics on the importance of the broad cooperative sector to the U.S. economy (Deller *et al.*, 2009). Also, in 2010, the Measuring the Cooperative Difference Research Network initiated a five-year project aimed at measuring the economic, along with social and environmental, impact of all cooperatives operating in Canada (Measuring the Cooperative Difference Research Network, 2010). As well, in 2012, Coop FR – les entreprises coopératives produced the first set of statistics on the contribution of the broad cooperative sector to the French economy (Coop FR – les en-

---

[1] Cooperatives succeed not only in a time of economic crisis, but also when the economy is good (see Mazzarol *et al.* (2014) for case studies of organizational resilience in the cooperative business model from around the world and across a wide range of industries; also, see Quebec Ministry of Economic Development, Innovation and Export (2008), Murray (2011), and Stringham and Lee (2011) for studies of the survival rates of cooperatives versus those of other business enterprises in various Canadian provinces).

treprises coopératives, 2012).[2] Prior to these initiatives, the analysis of the economic impact of cooperatives was generally limited to a province/state or regional economy[3] and/or to a single economic sector[4] (the one exception that we know of is Cooperatives UK, who started to produce a comprehensive review of the UK's cooperative sector in 2007 (Cooperatives UK, 2007)).

Economic impact is broadly defined as the economic benefits that a particular business or sector brings to the local, regional and national economies, and the communities in which it operates. In the cooperative literature, a distinction is often made between: (a) benefits to cooperative members – the so-called intra-cooperative value and (b) benefits to the larger community and the economy – the so-called extra-cooperative value (McKee et al., 2006). Table 1 presents examples of such benefits at the member versus community level.[5]

Table 1. *Examples of cooperative economic impact at the member versus community level*

Level	Examples of Economic Impact
Cooperative Member	Improved profitability as a result of better prices, lower costs, additional revenue from value added activity, and enhanced market access (e.g., USDA, 2001; Leclerc, 2010; London Economics, 2008; Zeuli and Deller, 2007)
	Access to goods and services not otherwise provided (e.g., IRECUS, 2012; Leclerc, 2010; Folsom, 2003; Fulton and Ketilson, 1992; Zeuli and Deller, 2007)
	Increased income as a result of patronage refunds (e.g., Deller et al., 2009; Zeuli et al., 2003; Cooperatives UK, 2014)

---

2   Recent work on the economic impact of cooperatives also includes IRECUS (2012) for the 300 largest cooperatives in the world, among other studies.

3   In Canada, see for instance Herman and Fulton (2001) for Saskatchewan. In the U.S., see for instance McNamara et al. (2001) for the Great Plains and Eastern Cornbelt, Folsom (2003) for Minnesota, Zeuli et al. (2003) for Wisconsin, and Coon and Leistritz (2005) for North Dakota.

4   For example, electric cooperatives in Iowa (Strategic Economics Group, 2006).

5   There are three types of impacts that need to be considered in economic impact studies – i.e., direct, indirect, and induced. The *direct* impacts are attributable to the activity of cooperatives and represent, for instance, the revenue generated by selling the output, the income paid to workers and owners (i.e., wages, salaries, and patronage refunds), the number of people employed, the amount of taxes paid, etc. The *indirect* impacts result from the activities of businesses that supply cooperatives with goods and services. Finally, the *induced* impacts result from the re-spending of member patronage refunds and worker income on consumer purchases.

	Business income/sales volume (e.g., Deller *et al.*, 2009; Folsom, 2003; Zeuli *et al.*, 2003; Ketilson *et al.*, 1998; Herman and Fulton, 2001; Bangsund *et al.*, 2011; Cooperatives UK, 2014)
	Capital investment/Assets (e.g., Ketilson *et al.*, 1998; Herman and Fulton, 2001)
	Value added/Gross Domestic Product (e.g., Zeuli *et al.*, 2003)
Community and Economy	Employment; wage and salary income (e.g., Deller *et al.*, 2009; Folsom, 2003; McNamara *et al.*, 2001; Zeuli *et al.*, 2003; Ketilson *et al.*, 1998; Herman and Fulton, 2001; Leclerc, 2010; Bangsund *et al.*, 2011; The ICA Group, 2012; Cooperatives UK, 2014)
	Government tax revenues (e.g., Deller *et al.*, 2009; Folsom, 2003; Zeuli *et al.*, 2003; Leclerc, 2010; Bangsund *et al.*, 2011)
	Pro-competitive effect on product price and quality, wages (e.g., USDA, 2001; Leclerc, 2010; The ICA Group, 2012; Zeuli and Deller, 2007)
	Economic stability (e.g., IRECUS, 2012; The ICA Group, 2012; Zeuli and Deller, 2007)

The following cooperative economic impact measures have been used most often in impact studies to date: (a) assets; (b) business sales; (c) membership; (d) patronage refunds; (e) wages and salaries; (f) jobs (employment), and (g) taxes. While there is consensus that the other benefits should also be included for a complete measure of the economic impact of cooperatives, they are more difficult to quantify and thus have been only discussed in economic impact studies to date (e.g., Fulton and Ketilson, 1992; Folsom, 2003; Leclerc, 2010; The ICA Group, 2012).

Economic impact studies of the broad co-operative sector are important for at least two reasons. First, studies of this nature help to understand the role co-operatives play in the economy and to determine whether co-operatives generate a different level of economic impact relative to alternative forms of business organizations. Such estimates of the business volume and economic impacts of the cooperative sector are useful to a number of stakeholders. For example, the estimates are useful to governmental agencies in gauging the performance of cooperatives as a source of economic growth and a tool for community economic development. Also, cooperative trade associations[6] can use these estimates to underscore the importance of their membership and to justify political support for investment in cooperative development. In a similar vein, leaders of

---

[6] Examples include the Canadian Worker Cooperative Federation, the National Rural Electric Cooperative Association in the U.S., or the European Association of Cooperative Banks.

rural communities, who are often involved in attracting or developing value-added cooperative businesses, can use these estimates to justify incentive packages. Second, conducting an economic impact study allows the co-operative sector to benchmark its performance across time, as well as relative to other sectors, to encourage improvement.

Conducting an economic impact study of the cooperative sector is not without challenges, and awareness of these challenges and ways in which they may be addressed can help the planning and actual implementation of the analysis, and can enhance the quality of the results. However, discussions of the topic in both the economic and cooperative literatures tend to revolve around determining the most appropriate method to be used for measuring economic impact in a certain context (e.g., Loveridge, 2004; Deller *et al.*, 2009; Leclerc, 2010; Uzea, 2014).[7] With the exception of McKee *et al.* (2006), there is little work done on other challenges that are likely to arise during the actual implementation of an economic impact study of the cooperative sector, to our knowledge.

The objective of this chapter is to start filling that gap. A review of published literature[8] on the economic impact of cooperatives is undertaken in order to identify potential challenges and provide preliminary insights into how they can be addressed.[9] As Figure 1 shows, these difficulties fall into three main categories, corresponding to the three stages of an impact analysis – that is, (1) data collection issues, such as obtaining access to existing microdata, standardizing economic activity data across existing datasets, enumerating the co-operative population, defining the time scale, defining the region of study, and surveying cooperatives; (2) data analysis issues, such as identifying the unit of analysis, adjusting the standard impact methods to suit the specifics of cooperatives, accounting for the unique outcomes of cooperatives, and examining the distribution of impacts, and (3) issues related to the interpretation of results, such as defining the counterfactual state and comparing the impact of cooperatives to that of the non-cooperative sector or across time. The next section discusses these challenges and the ways in which cooperative impact studies to date have dealt with them. The concluding section

---

[7] Economic impact methods differ in data requirements, assumptions, computational complexity and extent/nature of impacts captured, and model selection involves trade-offs. Moreover, all standard economic impact methods fail to account for the unique outcomes of cooperatives, such as competitive yardstick, missing goods and services, and local economic stability (Uzea, 2014).

[8] The literature review covered exclusively the studies published in English and French.

[9] The findings from this study directly inform the project led by the Measuring the Co-operative Difference Research Network (MCDRN) in Canada – the National Study on the Impact of Cooperatives (NSIC).

presents the key findings and puts forward a number of assertions and ideas for future research.

Figure 1. *Challenges in conducting an economic impact study of the cooperative sector*

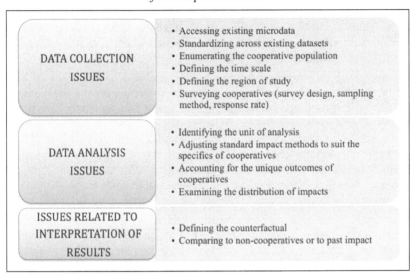

2. The challenges

2.1. Data collection issues

Conducting an economic impact study of the co-operative sector requires adequate economic activity data for the entire population of co-operatives. If this data is readily available for analysis, one can proceed to analyze it. However, this is almost never the case. Instead, the following three situations are more common: all needed data exists, but it is spread across multiple datasets; data does not exist and needs to be collected through survey, and existing data needs to be complemented with survey data.

2.1.1. All needed data exists, but it is spread across multiple datasets

One situation is when data has already been collected, but multiple existing datasets need to be compiled in order to get a full portrait of the cooperative economy. In Canada, for example, the Annual Survey of

Canadian Cooperatives conducted by Agriculture and Agri-Food Canada since 1931 is a rich data source (it has historically had a response rate close to 70%); however, it only surveys non-financial cooperatives. Thus, to get a full picture of the Canadian cooperative sector, this dataset had to be combined with data on the cooperative financial sector, which is housed at the Credit Union Central of Canada. Two issues are likely to arise when trying to access and compile existing datasets: obtaining access to microdata and standardizing economic activity data across highly heterogeneous datasets.

Accessing existing microdata. Obtaining access to microdata of existing datasets can be challenging for at least two reasons. First, it may be difficult, if not impossible, to search existing datasets for cooperative status. This issue was faced by Deller *et al.* (2009) when they attempted to search government data on property assessments and property taxes for housing cooperatives in the U.S. Second, confidentiality of economic activity data can pose problems in accessing existing microdata – for instance, when there are not many cooperatives in a sector or region.

Standardizing economic activity data across datasets. This step becomes problematic when there are differences in reporting practices across jurisdictions, economic sectors, or time, which lead to inconsistency between different datasets. For example, government data on property assessment and property taxes for housing cooperatives across the U.S. is not consistent since assessment and taxing practices vary across municipalities (Deller *et al.*, 2009). Also, Department of Justice data on cooperative membership in Saskatchewan is not consistent across time due to cooperatives switching from reporting total membership to only reporting active membership (Herman and Fulton, 2001). Another situation is when data on certain economic impact measures are missing altogether from one or more of the datasets that are to be compiled.

2.1.2. Data does not exist and needs to be collected through survey

Most of the time, however, the data does not exist and needs to be collected through surveys (as in Bhuyan and Leistritz, 1996; Coon and Leistritz, 2001, 2005; Zeuli *et al.*, 2003; Folsom, 2003; Deller *et al.*, 2009). This may pose a number of challenges, including: enumerating the co-operative population, defining the time scale, defining the region of study, and surveying cooperatives.

Enumerating the co-operative population. Determining the entire population of co-operative businesses is likely to be a complex task, particularly when done nationwide. While co-operatives operating in major sectors (e.g., agricultural marketing co-operatives or credit unions) have provincial/national trade associations and can be relatively easy to identify, other co-operatives have no industry-wide representation. It is these co-operatives that may be challenging to identify and, thus, missed from the analysis. For instance, the census of U.S. co-operatives conducted by Deller *et al.* (2009) identified more than 29,000 co-operatives operating in the U.S. economy, while the National Cooperative Business Association (NCBA, 2005) reported only 21,000 co-operatives, based on readily identifiable business sectors. Apart from lists maintained by trade associations, Deller *et al.* (2009) located co-operatives through academic collaborators, web searches, and searches of databases of non-profit organizations.

A related problem when trying to enumerate the co-operative population is distinguishing cooperative businesses from alternative business models. This is going to be a complex task as the cooperative business model itself is changing to ameliorate perceived financial constraints (e.g., Chaddad and Cook, 2004; Baarda, 2006).[10] Criteria for identifying co-operative businesses must be established. They must be broad enough to provide a comprehensive census of cooperative businesses, but not so broad as to lead to double-counting. A strategy on how to consistently use these criteria across sectors, jurisdictions, and time also needs to be developed.

To assemble the relevant population, most impact studies (e.g., Zeuli *et al.*, 2003; Karaphillis, 2012; Bhuyan and Leistritz, 1996; Coon and Leistritz, 2001, 2006; McKee, 2011; Folsom, 2003) have considered all organizations that have incorporated as cooperatives. However, as Deller *et al.* (2009) argue, businesses that function as cooperatives may be incorporated as limited liability companies, corporations, or non-profits. Thus, use of incorporation status may not provide a comprehensive cooperative census. For example, the National Study on the Impact of Co-operatives (NSIC) in Canada recently had to decide whether to include mutuals. While not incorporated as cooperatives, mutuals walk a fine cooperative line and have recently been fully embraced by the national apex organization for cooperatives.

Other potential criteria for identifying co-operatives include: application of co-operative principles, self-identification, tax filing status, and ownership and governance structure. Neither of these criteria is sufficient

---

[10] A wide variety of cooperative models are emerging, including new generation cooperatives, base capital plans, subsidiaries with partial public ownership, equity-seeking joint ventures, combined limited liability company-cooperative strategic alliances, and permanent capital equity plans, among others. See Chaddad and Cook (2004) for a discussion of the organizational attributes of these non-traditional cooperative models, including ownership structure, membership policy, voting rights, governance structure, residual claim rights and distribution of benefits.

on itself and they can sometimes be in conflict (Deller *et al.*, 2009). For instance, assessment of whether an organization has the characteristics of a cooperative and applies cooperative principles can be done by reading its bylaws and articles of incorporation; however, this is an impractical screening mechanism to build a census. Self-identification – the use of the term "cooperative" in the organization's name – is not a reliable indicator of the cooperative character of an organization, as some organizations operate as cooperatives but do not have "cooperative" in their name. The use of tax-filing forms cannot provide a comprehensive census of cooperative businesses; a business run on a cooperative basis may file a standard corporate income tax return when it has a large share of non-member business[11] or receives a large share of non-member equity capital to qualify for tax treatment as a cooperative.[12] Also, the tax-exempt status is not a reliable filter for identifying cooperatives, as cooperatives have tax-exempt status in sectors where non-cooperative, non-profit organizations also operate.

It is believed by Deller *et al.* (2009) that organizations can be reliably identified as cooperatives through a combination of incorporation status, tax filing or tax-exempt status, and member activity information (i.e., membership criteria, member voting rights for board elections, patronage refund allocation and non-participation on the board by management). An already complex task, identifying cooperatives is going to be even more difficult in those cases when an organization meets some but not all of the criteria.

Defining the time scale. Yet another decision must be made about the minimum length of time a co-operative must have conducted business before it is considered a member of the population. Findings from a study of the Western Areas Cities and Counties Cooperative (WACCO) – a cooperative of local governments in western Minnesota (U.S.) – revealed that "even well after operations are under way, the benefits of a cooperative can be difficult for many to perceive. WACCO, for example, has helped reduce costs and improve the quality of local government services, but it is essentially invisible to local citizens (USDA, 2001, pp. 81)." Similarly, a decision must be made about the maximum length of time since a change in business structure (e.g., demutualization) before firms are no

---

[11] This problem will not arise in those jurisdictions (e.g., Quebec, Canada) where legislation forces cooperatives to change their legal status when the share of non-member business exceeds 50%.

[12] A survey of businesses operating in biofuels, consumer goods, arts and crafts, and social and public services sectors in the U.S. revealed that: a) 15% of the sampled firms that incorporated as cooperatives chose to file as standard business and b) 26% of the sampled firms that incorporated as a corporation filed as cooperatives (Deller *et al.*, 2009).

longer considered part of the population. In other words, there should be some decision as to whether the full potential economic benefits have been obtained for business conducted within the study period.

Defining the region of study. When measuring local impact, a situation may arise in which some cooperatives operating in the area under study may be headquartered elsewhere, or they may be incorporated in that area but have affiliates or subsidiaries in another jurisdiction within the nation or internationally. Thus, a decision needs to be made on whether to include all co-operatives operating in a given area or only those co-operatives incorporated in that area. These issues were addressed by, for instance, Folsom (2003) and Zeuli *et al.* (2003) since Minnesota and Wisconsin are home to many large co-operatives that are headquartered elsewhere and/or have significant business out of state. Folsom (2003) included only the co-operatives incorporated in Minnesota and asked the co-operatives to estimate the percentage of expenditures going out-of-state. The author then used this percentage to reduce the gross-revenue figures provided by the co-operatives, acknowledging that the percentage may have been a rough estimate and thus may not have led to entirely accurate results. Zeuli *et al.* (2003) extended the analysis to some co-operatives operating in Wisconsin but incorporated in another state, rather than focusing only on those incorporated in the state. They asked those co-operatives operating in Wisconsin but headquartered elsewhere to provide gross sales, number of members and employees, and salary figures that represented only the Wisconsin portion of their business.

The choice between considering all cooperatives, regardless of where they are headquartered, and focusing only on those cooperatives headquartered in the area under study will likely be driven by two factors: (a) availability of reliable data at the establishment level, and (b) the tradeoff one is willing to make between the scope of the study and the quality of the data collected (and ultimately the results). When the scope of the analysis is restricted to cooperatives incorporated in the area under study, cooperative impact is nonetheless underestimated and the analyst will need to provide a sense of the extent of that underestimation. Economic studies to date at most specified the number of cooperatives that were left out of the study, with no estimate for the associated economic impact (e.g., McKee, 2011).

Measuring the economic impact of cooperatives at the national level should be easier than at the regional level as there is no need to quantify the state-specific level of business for cooperatives that do business in multiple states/provinces. Data will only need to be disaggregated at the domestic and international levels; however, since most cooperatives generally participate only in domestic commerce, the extent of disaggregation that needs to happen will be minimal.

Surveying cooperatives. Upon identification of the cooperative population, a survey would gather economic activity data such as revenues, employment, wages and benefits, direct expenditures, profit distribution, and tax payments. A cooperative specific survey could also collect cooperative specific data such as member shares' value and governance participation, as well as data regarding social and environmental impacts; however, care must be taken in requesting too much information, as it may lead to a low response rate. A standardized survey instrument and a uniform sampling methodology need to be used to minimize measurement error and yield data that would be comparable across economic sectors (Deller *et al.*, 2009). A useful sampling method is stratified sampling – i.e., in business sectors with relatively few cooperatives, all cooperatives are surveyed; in sectors with many cooperatives, a stratified random sample will be selected such that it represents the underlying distribution for the sector.

Like with any survey, conducting a survey of cooperative businesses will have to deal with how to achieve a sufficient response rate in order to make statistically valid statements about the population. Not unlike other social economy actors, many cooperatives, especially those of a more social nature (e.g., housing, health care, education or community services) are understaffed; therefore, accessing cooperative executives, board members or volunteers to fill in a survey could be difficult. Telephone interviews tend to be more successful than online, email, postal or fax surveys, as it allows interviewers to better convey the importance of the study and reassure respondents that the time commitment is minimal (London Economics, 2008). However, it is critical that appointments are scheduled in advance, to give respondents time to collect financial information before the actual telephone survey (Deller *et al.*, 2009). Nonetheless, it is important to extensively publicize the survey to increase participation. A wide range of mediums can be used to publicize the survey such as: (a) have cooperative trade associations distribute invitations to their member lists, on their websites, and in their newsletters; (b) mail and e-mail invitation letters, and (c) extend direct invitations by telephone (Deller *et al.*, 2009). If available, secondary data from cooperative financial reports or public databases can be used to complement survey information. However, this will likely lead to another issue, that of heterogeneity of data sources and reporting of financial information.

2.1.3. Existing data needs to be complemented with survey data

Finally, there may also be the case that data exist for some sectors or impact measures, but not for others. A survey will collect data on the latter sectors or impact measures, and the survey data will be combined with the existing data. For instance, in Zeuli *et al.* (2003) and Folsom (2003), survey data on cooperatives was combined with existing data on credit unions. This situation is the most complex as it poses both the challenges of collecting new data, and the challenges of accessing existing data and compiling different datasets. Depending on how significant the difficulties of accessing existing data and compiling datasets are, it may be more advantageous to collect new data for the sectors or impact measures for which data already exists.

2.2. Data analysis issues

The following challenges will arise during the data analysis stage: identifying the unit of analysis, adjusting the standard impact methods to suit the specifics of cooperatives; accounting for the unique outcomes of cooperatives, and examining the distribution of impacts.

Identifying the unit of analysis. It is important that the impact analysis is conducted at the sector and sub-sector level. Grouping all co-operatives together creates an aggregation bias (since co-operatives in different sectors have different product mixes, technology, and behavior) and may change the total impact numbers. It also hides the respective impacts of each co-operative type. However, the impact analysis by economic sector and sub-sector is challenging, as not all co-operative businesses fall completely into one sector or another (for example, while most cooperatives would fall under either the agricultural marketing or farm supply sector, the Hensal District Co-operative[13] in Canada generates revenue from both marketing of grains and supply of farm inputs). Thus, criteria must be established for splitting the data. In addition, as Deller *et al.* (2009) found, the reporting of financial information (e.g., definition of patronage refunds) tends to vary by sector which poses challenges to standardizing the data for analysis. Despite these challenges, most impact studies divided the cooperative population by major economic sectors. Only a handful of studies (e.g., Leclerc, 2010) analyzed all cooperatives together.

Adjusting the standard impact methods. The methodologies most commonly used to measure the contribution of a business sector to the local or national economy include: (a) the "head-count" approach; (b) the input-output (I-O) model; (c) the social accounting matrix (SAM) approach, and (d) the computable general equilibrium (CGE) model.[14] To briefly review, the "head-count" approach is used to assess the relative size of a sector by simply inventorying the assets held, the amount of

---

[13] www.hdc.on.ca/.
[14] Loveridge (2004) offers a summary and critical review (i.e., general operating principles, major shortcomings, and appropriate uses) of the I-O, SAM, and CGE models. Also, see Uzea (2014) for a detailed examination of the limitations of all these methods when measuring the economic impact of the cooperative sector.

capital investment made, the revenues and profits generated, the number of people employed, and the wages, salaries and dividends paid, among other indicators.

The I-O model starts with a transactions matrix that describes, in an equilibrium framework, the sales and purchases of goods and services between all sectors of the economy for a given period of time.[15] Algebraic manipulation of the model (i.e., matrix inversion) allows the modeler to determine the economy-wide impacts of an increase in the activity of a sector – *the multiplier effects*. The multipliers are then used to generate estimates for the *direct, indirect,* and *induced effects* of the change in economic activity. The main drawback of the I-O model is the assumption that resources flow freely to the industry under study and related industries (i.e., supplier industries). These resources are assumed to not be used elsewhere; hence, there is no reduction in output elsewhere or increase in input prices.

SAM models operate with the same basic set of assumptions and solution method as I-O models. However, a SAM is a more comprehensive database than the transactions matrix of an I-O model. Apart from the purchase or selling of goods and services, transactions in a SAM also include transactions that take place during the production process (e.g., the purchasing of intermediate goods and hiring of factors), current account transactions of institutions[16] (e.g., inter-institutional transfers and the payment of various taxes), capital account transactions of institutions (e.g., savings and investments), and any transaction that takes place across international borders (e.g., foreign direct investment and international trade transactions). Because the estimated impacts in a SAM model are broken down into finely disaggregated industries, SAM models are better suited when special consideration is given to the distributional aspects of an increase in the economic activity of a sector.

In a CGE model, each transaction flow in the SAM table is disaggregated into two components – price and quantity – which are allowed to adjust in response to the increase in the economic activity of the sector under study. Technically, a CGE model consists of a system of (a) simultaneous equations – i.e., supply and demand equations describing the behavior of economic agents, and (b) macroeconomic constraints – i.e., macroeconomic aggregates and balances, such as investments and savings, balance of payments, etc. CGE models can be static (i.e., no time dimension) or dynamic (i.e., explicitly consider time and time-related adjustments), and are solved using equilibrium computation – i.e.,

---

[15] For each economic sector, total sales must be equal to total expenditures.
[16] Institutions refer to households, businesses and government.

equilibrium is reached when a vector of prices is found that "clears"all the markets, while satisfying all the macroeconomic constraints. A typical CGE model gives a measure of the overall change in economic output through the effect on GDP, while also providing output results for individual industries. The impact on key variables such as employment or government revenue is also part of the model's output.

Table 2. *Review of economic impact studies of cooperatives*

Method	Author, year of publication	Scope of the study
"Head-Count" Approach	Fulton *et al.*, 1991	Various sectors, Saskatchewan
	Ketilson *et al.*, 1998	Various sectors, Saskatchewan
	Herman and Fulton, 2001 National Cooperative Business Association, 2005	Various sectors, Saskatchewan Various sectors, U.S.
	Cooperatives UK, 2007-2014 Coop FR, les entreprises coopératives, 2012, 2014	Various sectors, U.K. Various sectors, France
Input-Output Model	Bhuyan and Leistritz, 1996	Various sectors, North Dakota
	Bangsund and Leistritz, 1998	Sugar beet industry, North Dakota and Minnesota
	Bhuyan and Leistritz, 2000	Non-agricultural sectors, North Dakota
	McNamara *et al.*, 2001	Agriculture, Great Plains and Eastern Cornbelt
	Coon and Leistritz, 2001	Various sectors, North Dakota
	Folsom, 2003	Various sectors, Minnesota
	Coon and Leistritz, 2005	Various sectors, North Dakota
	Deller *et al.*, 2009	Various sectors, U.S.
	Leclerc, 2010	Various sectors, New Brunswick
	McKee, 2011 John Dunham & Associates, 2011 Frick *et al.*, 2012	Various sectors, North Dakota Food industry, Vermont Various sectors, Montana
	Karaphillis, 2012, 2014	Various sectors, Nova Scotia
Social Accounting Matrix Approach	Zeuli *et al.*, 2003	Various sectors, Wisconsin
General Equilibrium Model	N.A.	

The tradeoff between data requirements and computational complexity, and how well the model reflects reality have made I-O the most common tool for the measurement of the economic impact of co-operatives (see Table 2). However, it is important to note that I-O has a number of limitations when it comes to measuring the economic impact of co-operatives (as discussed in Zeuli and Deller, 2007).

Local purchasing. A key limitation when using I-O to measure the contribution of cooperatives to the local economy is the inability to account for the unique relationship cooperatives have with the local economy. Specifically, within the standard I-O tables, the multipliers are assumed to be the same for all business structures within a single industrial sector.[17] However, cooperative theory suggests that cooperatives are likely to purchase more of their inputs locally than other types of firms within the same industry classification (Fulton and Ketilson, 1992; Fairbairn *et al.*, 1995). Since cooperative owners are also community residents, they may support the purchase of local inputs (even if they are more expensive) because they will benefit from the long-term positive economic and social impacts that local businesses have on their community. Also, there is empirical evidence that cooperatives have competitive advantages at sourcing locally relative to other business structures (e.g., Enlow, Katchova, and Woods, 2011; Katchova and Woods, 2011).

When undertaking an economic impact study of the cooperative sector, a survey of purchasing patterns needs to be conducted to test the hypothesis that cooperatives purchase more locally than comparable firms with other business structures.[18] If spending patterns do differ among business structures,the I-O estimates are going to be biased. To get unbiased estimates, economic sectors based on standard system codes need to be further refined according to business structure (i.e., cooperative vs. non-cooperative) and new multipliers need to be calculated for the cooperative businesses.

Patronage refunds. The I-O model also has limitations when it comes to analyzing the impact of patronage refunds. Cooperatives mainly use patronage refunds to share net profits with their members, a different mechanism from the dividends used by investor-owned firms. However, the national firm surveys used to update the national I-O tables ask for information about dividends and not patronage refunds. Since patronage refunds are not exactly the same as dividends, these surveys do not provide accurate data on patronage refunds. Also, once the data is aggregated, the significance of patronage refunds gets lost, especially in those sectors where cooperatives represent a small share of the total firm population. If patronage refunds are not properly accounted for, the local economic

---

[17] There are a number of standard systems that are used to group economic activities into sectors, including: the North American Industry Classification System (NAICS), the International Standard Industrial Classification (ISIC) system of the United Nations, the Australian New Zealand Standard Industrial Classification (ANZSIC) system, and the UK Standard Industrial Classification (UKSIC) system, to name but a few.

[18] Food cooperatives in the U.S. have been found to purchase more locally than conventional retailers (The ICA Group, 2012).

impact of cooperatives is underestimated – that is, the value-added estimate for local economies is likely much higher for patronage refunds than dividends since cooperatives tend to be locally owned, unlike for-profit firms whose owners are generally spread across the country or internationally. Moreover, even when ownership of for-profit firms is local, it tends to be concentrated in the hands of a few people (e.g., family-owned businesses), unlike cooperatives which are owned by many people.

Even if patronage refund data is collected, there is also the question of how to analyze it. Total income within an I-O framework comprises personal income and property income, and dividends are included in property income. However, as Folsom (2003) notes, treating patronage refunds as property income is incorrect since they can be subject to different corporate level taxation rates than dividends.[19] Moreover, the assumption of the I-O model that some revenue leaks out of the region (to reflect returns to non-local investors) might also be inappropriate, as all of a cooperative's patronage refunds may be returned locally (Folsom (2003) assumes that 100% of spending stays local). In response to the patronage refund issue, Folsom (2003) chose to consider patronage refunds as part of personal income. In Zeuli *et al.* (2003), patronage refunds were treated as a separate influence to final demand, thereby creating their own set of impacts (in terms of total income and tax revenues).

Top-down versus bottom-up. When the I-O model is used to measure the economic impact at a local level, it is important to consider how well it reflects the region of study. Many models are created "top-down," where the national I-O model is used as a benchmark for building local models based on secondary data. The implicit assumption in these models is that local production technologies are identical to national averages. However, if the local economy is unique, with substantially different production technologies than national averages (e.g., reflecting the local availability of various goods or services), these models produce inaccurate results.

A more accurate analysis would require constructing a model "bottom-up" using survey data. That is, the analyst will have to manipulate the underlying structure of the I-O table to appropriately reflect the local economy through surveying a representative and sufficiently large sample of the local industries. The North Dakota Input-Output Model (Coon *et al.*, 1985; Coon *et al.*, 1989), which all North Dakota economic impact studies of the cooperative sector use, is built around primary survey data. However, such models are expensive and time consuming to build and update. A hybrid approach, which combines secondary data with survey

---

[19] For instance, the French law considers patronage refunds as personal income and not property income; hence, patronage refunds are subject to personal income tax rates.

data, can provide a balance between the accuracy of the table and the cost of constructing it (see Lahr (1993) for a discussion of the approaches and best practices that can be used to construct hybrid I-O models).

Accounting for the unique outcomes of cooperatives. It is important to recognize that all economic impact methods measure strictly the economic impact of cooperatives when viewed just like other business structures – i.e., measure benefits that accrue to the community, stakeholders, or the general public, such as tax payments or employment opportunities. They cannot assess the unique value of cooperatives, such as competitive yardstick in the market, and provider of valuable goods and services that would otherwise go missing in local communities. Moreover, by presenting a 'snapshot picture' of the economy, most of the methods (the dynamic CGE model being the exception) fail to consider the contribution of cooperatives to the long-term growth and resilience of the communities in which they operate. Previous studies have at best included a discussion of these economic impacts unique to cooperatives (e.g., Zeuli *et al.*, 2003; Leclerc, 2010; McKee, 2011; The ICA Group, 2012). Failure to quantify these "deeper impacts" (Deller *et al.*, 2009) understates the economic impact of the cooperative sector at the regional level, with this understatement even larger if omitted at the national level. Thus, additional analyses need to be completed if a more accurate assessment of the total value created by cooperatives is to be gained.

Pro-competitive effect. Cooperatives are typically seen as being pro-competitive market instruments (e.g., Nourse, [1922] 1992; Helmberger, 1964; Sexton, 1990; Innes and Sexton, 1994). The basic idea is that cooperatives provide "extra competition" that forces for-profit firms to operate in a more competitive way. Cooperatives may exert their competitive influence by, for instance, holding down prices of farm supplies and consumer goods and services, raising prices received for agricultural commodities, improving product quality, or raising industry standards for wage and benefits.

Although it seems to be a generally accepted view that cooperatives have a pro-competitive effect on the market, there is very little empirical evidence for the existence and magnitude of this effect. For instance, Jardine *et al.* (2014) is one of the few studies that confirm empirically the role of cooperatives in improving product quality. As regards the cooperatives' pro-competitive effect on prices, there is more evidence to draw on; however, the evidence is mixed. Fulton (1989), in an analysis of the fertilizer industry in western Canada, showed that cooperatives may fail to fulfill their pro-competitive role for various reasons, including barriers to entry by an incumbent for-profit firm. Also, Hoffman and Royer (1997) showed that the yardstick effect is not universal, but instead is sensitive to market structure and the behavior of cooperative members. In contrast, empirical evidence by Rogers and Petraglia (1994) and Zhang *et al.* (2007) supports the yardstick hypothesis.

Economic impact studies of the cooperative sector need to: (1) confirm existence of the pro-competitive effect of cooperatives and (2) estimate the impact it has on member (and non-member) welfare – e.g., the additional revenue or savings that members (and non-members) achieve as a result of the cooperative's presence on the market.

Missing goods and services. Cooperatives often operate in small local markets where investor-owned firms do not find it advantageous to do so. As Fulton and Ketilson (1992) show formally, this is due to the fact that the benefits to consumers from being able to purchase locally are greater than the business costs of supplying the good or service. Since, in cooperatives, unlike for-profit firms, these costs and benefits are incurred by the same group of people (i.e., the members), cooperatives will continue to operate in markets that do not allow for competitive rate of returns, providing valuable goods and services to the local community.

Economic impact studies of the cooperative sector need to: (1) establish whether existing cooperatives were created in reaction to either a for-profit firm leaving the community or a community need that was not being met by other businesses, and (2) measure the extra-benefit to cooperative members (and non-members) of having access to needed goods and services locally. Studies such as Fulton and Ketilson (1992) in Canada and Bhuyan and Leistritz (2000) in the U.S. provided evidence of cooperative businesses that were formed because for-profit firms were unable or unwilling to provide the goods and services. However, they did not proceed to measure the extra-benefit of the cooperative providing goods and services in the local communities that otherwise would not be provided.

Local economic stability. By the nature of their mission to serve members, cooperatives tend to be 'anchored' to the areas where members are located (Fairbairn *et al.*, 1995). This necessity to remain in proximity to members makes it less likely for cooperatives to relocate to places that might have cheaper raw products or labor, as is the case with for-profit firms. A related argument for the contribution of cooperatives to local economic stability is the fact that cooperatives have significantly higher survival rates than for-profit firms, at least based on evidence from developed countries (e.g., Murray, 2011; Stringham and Lee, 2011; Quebec Ministry of Economic Development, Innovation and Export, 2008).[20]

---

[20] For example, data from a 2008 study in Québec (Canada) shows that 62% of cooperatives survived after 5 years and 44% survived after 10 years. This compares to survival rates of only 35% after 5 years and 20% after 10 years for other forms of business organizations (Quebec Ministry of Economic Development, Innovation and Export, 2008).

Finally, cooperatives can also play a role in attracting and retaining additional economic activity (other cooperatives or for-profit firms) in a local area via, for instance, a healthy industry (Zeuli and Deller, 2007).

The contribution of cooperatives to local economic stability or long-term growth and resilience is perhaps the most important measure of their economic impact. Yet, most economic impact methods (dynamic CGE model being the exception) present a 'snapshot picture' of the economy, capturing the impact of cooperatives at a single point in time. To gauge the effect of cooperatives on community resiliency, one could compare economic indicators from a set of similar communities that differ only in the number and/or strength of cooperatives over a long-enough period of time. Alternatively, one could analyze the ability of a group of comparable communities (again allowing for differences in co-operation) to adjust to a similar local economic crisis (e.g., the loss of a major employer or industry) or to the recent global economic crisis.

Examining the distribution of impacts. In a time of economic stagnation, knowing the total economic impacts of the cooperative sector is particularly valuable. However, it would be worth analyzing, for instance, how the impacts are distributed across different sectors which were hit differently by the economic downturn. Similarly, it would be useful to know how the impacts are distributed between the less-advantaged rural areas and urban areas, as well as the type of jobs supported by cooperatives (part-time versus full-time, skilled versus unskilled).

The data requirements to undertake such analyses may become prohibitive though, as suggested by the small number of cooperative impact studies that have broken-down the impacts. Zeuli *et al.* (2003) and John Dunham & Associates (2011) are among the few studies that examined the distribution of the total economic impact of cooperatives across the different sectors, including agriculture, manufacturing, transportation, construction, services, and government, among others. Also, Coon and Leistritz (2001) differentiated between full-time and part-time jobs when analyzing cooperatives' direct impact on employment.

2.3. Issues related to the interpretation of results

The following challenges will arise during this stage: constructing the counterfactual state and comparing to non-cooperative sector or to past impact.

Constructing the counterfactual state. The main challenge when it comes to interpreting the results is defining the counterfactual. Computing the economic impact of co-operatives answers the question of what contribution the activity of cooperatives makes to the economy and involves a comparison of the current (observed) state of the economy with the counterfactual (hypothetical) state – that is, the state the economy would have been in but for the presence of cooperatives (Figure 2). While the current state of the economy can be assessed based on objective data, the counterfactual is defined by a theoretical model of the economy. The analyst uses a model of the economy to make judgments about, for instance, whether the output co-operatives add to the economy would have been provided by other business models, or the co-operative provides services that would not have otherwise existed. Similarly, whether the people co-operatives employ would be employed by other local firms, or would have to move out of the province or not work.

Figure 2. *Counterfactual Analysis*

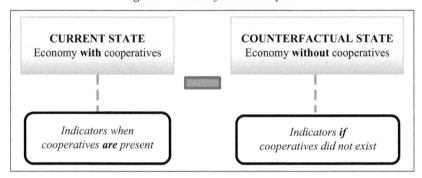

The implicit assumption in most of the economic impact studies of the cooperative sector is that the product and services would neither be provided nor exist in the economy if cooperative businesses did not exist (the same assumption is made with regards to employment). However, care must be made in considering how these benefits would not have otherwise occurred. To construct the counterfactual alternative, McNamara *et al.* (2001) asked co-operative managers to estimate what local employment and business impact would be felt by the local economy if the co-operative were to go out of operation. The authors found that higher population densities and greater economic diversification in rural Indiana counties lessens the potential impact of employment and business losses compared to the Colorado economy. Bangsund and Leistritz (1998), the other study that considered the counterfactual, showed that, in certain situations, multiple industries must be included in assessing the economic impact of co-operatives. Specifically, the study linked the impacts of North Dakota and Minnesota sugar beet production and processing industries, for without each other neither would exist (processing is completely owned by co-operatives). Due to its subjective nature ("a model is in the mind" as Heckman (2005, pp. 2) argues), defining the counterfactual is always going to be a controversial issue. Only robustness checks can be made with different hypotheses or sets of counterfactual situations.

Comparing the economic impact of cooperatives to that of the non-cooperative sector or comparing cooperative impact across time. An alternative approach to interpreting the results would be to compare the economic impact of cooperatives to that of other business structures (e.g., compare the impact of the financial cooperative sector to that of the private financial sector). However, this would significantly increase the data requirements, as the same data would need to be collected for the non-cooperative sector. While this could be less of a problem in those cases when the non-cooperative sector has done its own impact study, using those results requires that there is comparability between the studies in terms of data gathering and reporting, method used to analyze economic impact, time period and impacts covered, etc. We are not aware of any cooperative impact study that has done such a comparison.

Similar challenges will arise when trying to compare the economic impact of cooperatives over time. The need for consistency in terms of data gathering, reporting, and analysis was acknowledged by Herman and Fulton (2001) in a repeat study of the economic impact of cooperatives in Saskatchewan – the third[21] of its kind generated by the University of Saskatchewan Centre for the Study of Cooperatives: "To enable cross-study comparisons, we have endeavoured to duplicate the methodology of the previous studies (pp. 39)." Also, note that economic impact measures, such as revenues, profits and wages, were adjusted for inflation to allow comparison between different points in time.

3. Concluding discussion

This chapter aimed to identify the challenges to conducting a study on the economic impact of the broad co-operative sector, with special emphasis on national-level studies, and to provide preliminary insights into how they can be addressed. As presented, these challenges fall into three main categories, corresponding to the three stages of an impact analysis – that is, data collection issues, data analysis issues, and issues related to the interpretation of results. This section summarizes the findings, and puts forward some assertions and ideas for future research.

A key finding from the analysis in this chapter is the lack of economic activity data collected and compiled about co-operatives. Very few governments, apex organizations or co-operative sectors collect and compile

---

[21] After Fulton *et al.* (1991) and Ketilson *et al.* (1998).

co-operative economic data in a central, accessible, and reliable database. Without data that is standardized and accessible, conducting an economic impact assessment of the cooperative sector is time consuming at best and virtually impossible at worst – hence, the small number of studies. The importance of reliable co-operative economic activity data cannot go understated and the authors of this chapter encourage this task to be taken up by an independent body adept in data collection to avoid the pitfalls and challenges that have been discussed, including enumerating the co-operative population, defining the time scale, defining the region of study, and surveying cooperatives.

Another important finding is that standard economic impact methods (e.g., "head-count" approach, input-output model, social accounting matrix approach, computable general equilibrium model) cannot accurately measure the economic impact of cooperatives. The input-output model – the most common tool for economic impact analysis – has limitations when it comes to analyzing the impact of patronage refunds or accounting for the fact that cooperatives likely purchase more of their inputs locally than comparable firms with other business structures. Moreover, most methods measure strictly the economic impact of cooperatives when viewed just like other business structures and cannot assess the unique value of cooperatives for their members and their communities, such as competitive yardstick, missing goods and services, and local economic stability.

Some of these issues could be better addressed with a dynamic general equilibrium model, yet the need for additional data variables and computational complexity makes it undoable at this point for most cooperative sectors. But this does not mean researchers and the cooperative sector cannot strive for a "cooperative sector dynamic general equilibrium model" that can respond to the limitations discussed of the other methodologies. In the meantime, the input-output model can provide a good tradeoff between data requirements and computational complexity, and how well the model reflects reality. However, adjustments (discussed in this chapter) need to be made in order to be able to deal with the issues of local purchasing and patronage refunds. Moreover, the input-output analysis needs to be complemented with analyses of the competitive yardstick effect of cooperatives, the value of goods and services cooperatives provide that would otherwise go missing in local communities, and the contribution to local economic stability.

Finally, when it comes to interpreting the results from an economic impact study of the cooperative sector, the most important issue is defining the counterfactual state – what assumptions can be made about the economy if cooperatives were not present? Given its subjective nature, constructing the counterfactual is always going to be a controversial issue and only robustness checks can be made employing different hypotheses or sets of counterfactual situations. Alternatively, the economic impact of cooperatives can be compared to that of the non-cooperative sector.

This chapter is a small contribution to understanding the challenges researchers are up against when conducting economic impact analysis in the cooperative sector. Clearly, the scope for further research in this field is immense. As more impact studies are conducted in the co-operative sector, researchers can develop an understanding of how to best work with and around these challenges, and pave a standard way to best conduct cooperative economic impact studies. Future research could examine how to expand existing impact methodologies to capture the impact of competitive yardstick, missing goods and services, and contribution to local economic stability, in order to grasp the widest possible and actual economic impact of cooperatives. And there is an opportunity to expand impact methodologies to better be able to measure the social and/or environmental impact of co-operatives. Having a comprehensive understanding of the economic impact of co-operatives hand in hand with the social and environmental impact would paint a more complete picture of the co-operative landscape.

References

Baarda, J. R., *Current Issues in Cooperative Finance and Governance: Background and Discussion Paper*, Washington, D.C., United State Department of Agriculture, Rural Development, Cooperative Programs, April 2006.

Bangsund, D. A. and Leistritz, F. L., *Economic Contribution of the Sugar Beet Industry to North Dakota and Minnesota*, Research Report 395, Fargo, ND, North Dakota State University, Department of Agricultural Economics, February 1998.

Bangsund, D. A., Olson, F. and Leistritz, F. L., *Economic Contribution of the Soybean Industry to the North Dakota Economy*, Research Report 678, Fargo, ND, North Dakota State University, Department of Agricultural Economics, Agricultural Experiment Station, January 2011.

Birchall, J., *Resilience in a Downturn: The Power of Financial Cooperatives*, Geneva, International Labor Organization, 2013.

Birchall, J. and Ketilson, L., *Responses to the Global Economic Crisis: Resilience of the Cooperative Business Model in Times of Crisis*, Geneva, International Labor Organization, 2009.

Bhuyan, S. and Leistritz, F. L., *Economic Impacts of Cooperatives in North Dakota*, Research Report AE96009, Fargo, ND, North Dakota State University, Quentin Burdick Centre for Cooperatives and Department of Agricultural Economics, December 1996.

——, "Cooperatives in Non-Agricultural Sectors: Examining a Potential Community Development Tool," *Journal of the Community Development Society*, Vol. 31, 2000, pp. 89-109.

Chaddad, F. R. and Cook, M. L., "Understanding New Cooperative Models: An Ownership-Control Rights Typology," *Review of Agricultural Economics*, Vol. 26, No. 3, 2004, pp. 348-360.

Coon, R. C. and Leistritz, F. L., *Economic Contribution North Dakota Cooperatives Make to the State Economy*, Research Report AE01002, Fargo, ND, North Dakota State University, Department of Agribusiness and Applied Economics, March 2001.

Coon, R. C. and Leistritz, F. L., *Economic Contribution North Dakota Cooperatives Make to the State Economy*, Research Report AAE05001, Fargo, ND, North Dakota State University, Department of Agribusiness and Applied Economics, January 2005.

Coon, R. C., Leistritz, F. L., Hertsgaard, T. A., and Leholm, A. G., *The North Dakota Input-Output Model: A Tool for Analyzing Economic Linkages*, Agricultural Economics Report No. 187, Fargo, ND, North Dakota State University, Department of Agricultural Economics, 1985.

Coon, R. C., Leistritz, F. L. and Hertsgaard, T. A., *North Dakota Input-Output Economic Projection Model (NDIO/EPM): Documentation and User's Guide*, Agricultural Economics Software Series No. 4, Fargo, ND, North Dakota State University, Department of Agricultural Economics, 1989.

Coop FR, *Les entreprises coopératives (2012, 2014), Panorama sectoriel des entreprises coopératives et Top 100*, Paris, France, Coop FR, 2014.

Cooperatives UK, *Cooperative Review*, Manchester, UK, 2007-2009.

Cooperatives UK, *The UK Cooperative Economy 2013*, Manchester, UK, 2010-2014.

Deller, S., Hoyt, A., Hueth, B. and Sundaram-Stukel, R., *Research on the Economic Impact of Cooperatives*, Madison, WI, University of Wisconsin Centre for Cooperatives, March 2009.

Enlow, S. J., Katchova, A. L. and Woods, T. A., *The Role of Food Cooperatives in Local Food Networks*, Presented at the International Food and Agribusiness Management Association Annual World Forum and Symposium, Frankfurt, Germany, June 20-23, 2011.

Fairbairn, B., Bold, J., Fulton, M., Ketilson, L. and Ish, D., *Cooperatives and Community Development: Economics in Social Perspective*, Saskatoon, SK, University of Saskatchewan, Centre for the Study of Cooperatives, 1995.

Folsom, J., *Measuring the Economic Impact of Cooperatives in Minnesota*, RBS Research Report 200, Washington, DC, United States Department of Agriculture, Rural Business-Cooperative Service, December 2003.

Frick, M., Sheehy, J. and Nedanov, A., *Economic Contribution Montana Cooperatives Make to the State Economy*, Bozeman, MT, Montana State University, Division of Agricultural Education, February 15, 2012.

Fulton, M. E., "Cooperatives in Oligopolistic Industries: The Western Canadian Fertilizer Industry," *Journal of Agricultural Co-operation*, No. 4, 1989, pp. 1-19.

Fulton, M. E., Ketilson, L. and Simbandumwe, L., *Economic Impact Analysis of the Cooperative Sector in Saskatchewan*, Saskatoon, SK, University of Saskatchewan, Centre for the Study of Cooperatives, 1991.

Fulton, M. E. and Ketilson, L., "The Role of Cooperatives in Communities: Examples from Saskatchewan," *Journal of Agricultural Co-operation*, No. 7, 1992, pp. 15-42.

Heckman, J., "The Scientific Model of Casuality," *Sociological Methodology*, Vol. 35, No. 7, 2005, pp. 1-97.

Helmberger, P. G., "Cooperative Enterprise as a Structural Dimension of Farm Markets," *Journal of Farm Economics*, Vol. 46, No. 3, 1964, pp. 603-617.

Herman, R. and Fulton, M. E., *An Economic Impact Analysis of the Cooperative Sector in Saskatchewan: Update 1998*, Saskatoon, SK, University of Saskatchewan, Centre for the Study of Cooperatives, 2001.

Hoffman, S. H. and Royer, J. S., *Evaluating the Competitive Yardstick Effect of Cooperatives on Imperfect Markets: A Simulation Analysis*, Presented at the Western Agricultural Economics Association Annual Meeting, Reno/Sparks, Nevada, July 13-16, 1997.

Innes, R. and Sexton, R. J., "Strategic Buyers and Exclusionary Contracts," *The American Economic Review*, Vol. 84, No. 3, 1994, pp. 566-584.

Institut de recherche et d'éducation pour les coopératives et les mutuelles de l'Université de Sherbrooke (IRECUS), *The Socio-Economic Impact of Cooperatives and Mutuals*. Presented at the 2012 International Summit of Cooperatives, Québec City, October 8-11, 2012.

Jardine, S. L., Lin, C.-Y. and Sanchirico, J. N., "Measuring Benefits from a Marketing Cooperative in the Copper River Fishery," *American Journal of Agricultural Economics*, Vol. 96, No. 4, 2014, pp. 1084-1101.

John Dunham and Associates, *Vermont Food Industry Economic Impact Study*, Prepared for the Vermont Grocers' Association, New York, 2011.

Karaphillis, G., *Economic Impact of the Cooperative Sector in Nova Scotia*, Interim Summary Report presented at the Conference to Celebrate the International Year of Cooperatives, Halifax, NS, November 20-22, 2012.

Karaphillis, G., Lake, A. and Duguid, F., *Economic Impact of the Co-operative Sector in Canada*. Measuring the Co-operative Difference Research Network Working Paper, 2014.

Katchova, A. L. and Woods, T. A., *Local Food Procurement and Promotion Strategies of Food Cooperatives*, Paper presented at the Southern Agricultural Economics Association Annual Meeting, Corpus Christi, Texas, February 6-9, 2011.

Ketilson, L., Gertler, M., Fulton, M., Dobson, R. and Polsom, L., *The Social and Economic Importance of the Cooperative Sector in Saskatchewan*, Saskatoon, SK, University of Saskatchewan, Centre for the Study of Cooperatives, June 1998.

Lahr, M. I., "A Review of the Literature Supporting the Hybrid Approach to Constructing Regional Input-Output Models," *Economic Systems Research*, Vol. 5, No. 3, 1993, pp. 277-293.

Leclerc, A., *The Socioeconomic Impact of the Cooperative Sector in New Brunswick*, Moncton NB, University of Moncton, June 2010.

London Economics, *Study on the Impact of Cooperative Groups on the Competitiveness of their Craft and Small Enterprise Members*, Final Report to European Commission D.G. Enterprise and Industry, January 2008.

Loveridge, S. A., "Typology and Assessment of Multi-Sector Regional Economic Impact Models," *Regional Studies*, Vol. 38, No. 3, 2004, pp. 305-317.

Mazzarol, T., Reboud, S., Mamouni Limnios, E. and Clark, D., *Research Handbook on Sustainable Cooperative Enterprise: Case Studies of Organizational Resilience in the Cooperative Business Model*, Cheltenham, UK, Edward Elgar, 2014.

McKee, G. J., *The Economic Contribution of North Dakota Cooperatives to the North Dakota State Economy*, Fargo, ND, Department of Agribusiness and Applied Economics, North Dakota State University, Research Report No. 687, October 2011.

McKee, G. J., Kenkel, P. and Henehan, B. M., *Challenges in Measuring the Economic Impact of Cooperatives*, Presented at the NCERA-194 Annual Meeting, Minneapolis, MN, November 2-3, 2006.

McNamara, K. T., Fulton, J. and Hine, S., *The Economic Impacts Associated with Locally-Owned Agricultural Cooperatives: A Comparison of the Great Plains and the Eastern Cornbelt*, Presented at the NCR-194 Research on Cooperatives Annual Meeting, Las Vegas, NV, October 30, 2001.

Measuring the Cooperative Difference Research Network, "A Big Boost for Co-op Research," *Network Newsletter*, Vol. 1, No. 1, July 2010, p. 1.

Murray, C., *Co-op Survival Rates in British Columbia*, Canadian Centre for Community Renewal on behalf of the BC-Alberta Social Economy Research Alliance and British Columbia Cooperative Association, June 2011.

National Cooperative Business Association, *Cooperative Businesses in the United States: A 2005 Snapshot*, Washington, D.C., October 2005.

Nourse, E. G., "The Place of the Cooperative in Our National Economy," Reprinted in *Journal of Agricultural Co-operation*, Vol. 7, [1922] 1992, pp. 105-114.

Quebec Ministry of Economic Development, Innovation and Export, *Survival Rate of Cooperatives in Québec: Report Summary*, translated by the Ontario Co-operative Association, October 2008.

Roelants, B., Dovgan, D., Eum, H. and Terrasi, E., *The Resilience of the Cooperative Model: How Worker Cooperatives, Social Cooperatives, and Other Worker-Owned Enterprises Respond to the Crisis and Its Consequences*, CECOP-CICOPA Europe, June 2012.

Rogers, R. T. and Petraglia, L. M., "Agricultural Cooperatives and Market Performance in Food Manufacturing," *Journal of Agricultural Co-operation*, No. 9, 1994, pp. 1-12.

Sexton, R. J., "Imperfect Competition in Agricultural Markets and the Role of Cooperatives: A Spatial Analysis," *American Journal of Agricultural Economics*, Vol. 72, No. 3, 1990, pp. 709-720.

Strategic Economics Group, *Economic Impact Study of Iowa's Electric Cooperatives*, Prepared for the Iowa Association of Electric Cooperatives, Des Moines, Iowa, 2006.

Stringham, R. and Lee, C., *Co-op Survival Rates in Alberta*, Canadian Centre for Community Renewal on behalf of the BC-Alberta Social Economy Research Alliance and Alberta Community and Cooperative Association, August 2011.

The ICA Group, *Healthy Foods, Healthy Communities: Measuring the Social and Environmental Impact of Food Co-ops*, 2012.

United States Department of Agriculture, *The Impact of New Generation Cooperatives on Their Communities*, Rural Business Cooperative Service, RBC Research Report 177, 2001.

Uzea, N., *Methodologies to Measure the Economic Impact of Cooperatives: A Critical Review*, Measuring the Co-operative Difference Research Network Working Paper, 2014.

Zeuli, K. and Deller, S., "Measuring the Local Economic Impact of Cooperatives," *Journal of Rural Co-operation*, Vol. 35, No. 1, 2007, pp. 1-17.

Zeuli, K., Lawless, G., Deller, S., Cropp, R. and Hughes W., *Measuring the Economic Impact of Cooperatives: Results from Wisconsin*, United States Department of Agriculture, Rural Business-Cooperative Service, RBS Research Report 196, August 2003.

Zhang, J., Goddard, E. and Lehrol M., "Estimating Pricing Games in the Wheat-Handling Market in Saskatchewan: The Role of a Major Cooperative," *Advances in the Economic Analysis of Participatory and Labour-Managed Firms*, No. 10, 2007, pp. 151-182.

Mapping Social Enterprise in the UK

Definitions, Typologies and Hybrids

Roger SPEAR

Professor, Department of Innovation and Engineering,
Open University, United Kingdom

Introduction

Hybridity refers to a mixing of features from different types. There are different approaches to hybridity in the third sector and social economy, where so-called "social enterprises" seem to combine elements of non-profits, for profits, and co-operatives, and have caught the imagination of entrepreneurs, academics and policy makers alike.

The chapter examines how challenges arise in mapping social enterprise considered as a hybrid form, both conceptual and methodological. These include difficulties in defining operational criteria, matching to varied sampling frames, and making judgments about boundary cases. The chapter examines UK attempts to address these challenges but draws on other national and international experiences.

There are four major issues that need addressing for a successful mapping exercise: firstly definitional issues since the criteria for defining a particular type of organization are not always easy to apply; secondly population issues, since it is important to ascertain in which population of organizations can one find the defined type; thirdly the sampling frame issues (and sampling) since it is necessary to find one or more databases which includes the population of organizations of interest (even if other types are also included); and mapping strategies which may be several to triangulate the identification of the type of organization of interest.

Clearly definitional and population issues are closely interlinked, and these are both linked to sampling issues. This chapter draws on other experience of mapping non-profits and the social economy in order to get a greater understanding of mapping social enterprise sector. Thus for example it draws on elements of Johns Hopkins approach as a basis for addressing definitional issues, and making progress towards defining the social enterprise sector, and discussing the issues of boundary cases and hybridity.

In the next section we introduce the two core concepts, hybridity and boundary cases (section 1), which are then used to discuss the mapping of voluntary and non-profit sector (section 2), and social enterprises (section 3) in the UK. We conclude with comments on how some measures could improve the operationalization of the UK definition of social enterprise and help mapping the sector.

1. Hybridity and boundary cases, two operational concepts for mapping social enterprises

The social enterprise field shares with the social economy several different organizational forms, and growing tendencies towards hybridity both within the field and across its boundaries with the state sector, for-profit sector, and the community; this substantially exacerbates the difficulties of mapping this field.

1.1. Hybridity

Hybridity is a current theme in many fields of social science (e.g. Brah and Coombs, 2000). In studies of hybrid organizations, there are several different approaches. Most adopt approaches which define hybridity as the mixing of characteristics from different, adjoining sectors: public, business, third sectors. In the US, since Young's early work (1983) on entrepreneurial and commercial nonprofits, there has been a focus on the hybrid space between the nonprofit and business sectors. Battilana and Lee (2014) adopt a similar focus when conceptualizing social enterprise as a hybrid between business and charity, after a very extensive review of the literature (256 journal articles). They differentiate between three different types of hybridity: combining multiple organizational identities, combining multiple organizational forms, and combining multiple institutional logics. And based on their literature review, they identify five dimensions of hybrid organizing: inter-organizational relationships, culture, organizational design, workforce composition, organizational activities; they note that these dimensions vary across the spectrum between the social and the commercial, and that social enterprise may manage these differences through separation or integration.

A second approach with a similar focus on the nonprofit or voluntary organization, is that of Billis (2012) who develops a "prime sector" (ideal typical) approach, where hybrids are adaptations from this ideal. But he sees a greater plurality of hybrids adopting characteristics of adjoining sectors (public, private, third); and sometimes adapting so that they transition into the overlapping zones of different sectors. Key dimensions of the ideal types for all three sectors are: ownership, governance, operational priorities, human resources, and other resources. Changes in these dimensions towards the values of another sector indicates a hybridization process, with ideal types from different sectors transitioning towards another sector, and resting in a zone of hybridity between two sectors.

The third approach to hybridity, includes Evers (2005), and Evers and Laville (2004), who argue that hybridity is a result of "increasing intertwining of components and rationales," and that the third sector hybridity draws on a mix of resources, has a mix of goals, and "steering mechanisms" (governance principles) from three adjoining sectors of the welfare triangle: the market, the state and civil society sectors.[1] This broader perspective leads to multi-sectoral hybridity.

This latter perspective fits well with the solidarity economy (or new social economy), which is based on a Polanyian view of the economy, and thus moves away from the organizational view of the social economy based on statutes: Co-operatives, Mutuals, Associations/Non-profits, Foundations – in short CMAF. The solidarity economy is based on three interacting economic dynamics: redistribution of state resources, egalitarian reciprocity, and market exchange; but it is also informed by a strong associational dynamic which emphasizes that solidarity economy also operates at the social and political level where it serves to coordinate citizen democratic action, through solidaristic relations and public authority. Thus this perspective challenges conventional categories of the economy, and of organization, and provides a new understanding of hybrids and the processes that drives them – so for example hybrids are formed around a more permeable informal/formal boundary of organization.

With regard to the growing prominence of the social enterprise hybrid, Bode, Evers and Schulz (2006) regard this as having a similar set of distinctive features (p. 237): multiple goals, multiple resources, and multi-stakeholders. Similar positions are established by Brandsen, van de Donk, and Putters (2005), and these various authors argue that hybridity and change are permanent features of third sector and social/solidarity economy, as they struggle to manage tensions between institutional logics and logics of provision (e.g. Lounsbury and Boxenbaum, 2013).

Spear (2012) also develops an ideal typical approach to analyze hybridity of the co-operatives sector, based on a number of different

---

[1] NB civil society includes third sector organizations, and has growing importance given the reformulation of policy thinking about the role of volunteers and family in service provision.

tendencies including changes to: membership, finance, managerial control, and governance. This represents a pattern of hybridization in one organizational form within the social economy, and most patterns of hybridization identified are towards the business sector. But one form, the social co-operative is a hybrid *within* the social economy, between co-operatives and non-profit associations.

Although there may also be isomorphic tendencies within the social economy – thus for example different types of third sector/social economy structures (CMAF) may be subject to institutional processes which reinforce their distinct and separate identities. But more frequently, after phases of hybridization, there may be episodes of institutionalization (e.g. legislation) which consolidate the new hybrid form. For example in Italy legislation and policy frameworks for social co-operatives (hybrids between non-profits and co-operatives) was established in 1991, many years after the initial hybridization and after considerable numbers had been formed. The EMES network has generally adopted a similar *intra-sectoral* approach to social enterprise hybrids, seeing them as a mix of nonprofits and co-operatives (Borzaga and Defourny, 2001).

But hybridity (of social enterprise) is more commonly seen as inter-sectoral, thus in some countries social enterprise may combine features of non-profits and for-profits – again in Italy, legislation for social enterprises was established in 1996 – this was a broader hybrid form allowing any form of legal structure to qualify as a social enterprise including commercial business structures providing they meet other criteria; a similar qualifying legal status was used in the UK for its Community Interest Company which allowed share-based structures and third sector structures to qualify as social enterprise, albeit with constraints on features like profit distribution.

The hybridizing process has also been the subject of study. Borzaga and Defourny (2001) argue that social enterprise are formed both by changing contexts (such as increased marketization of welfare service provision) influencing established organizations, as well as new social enterprise being formed. Battilana and Lee (2014) also note hybridization developing along a continuum between intersectoral types, rather than a dichotomous formation process where the boundary between the hybrid and its sectoral ideal types is clear. They differentiate between hybridity at the core and periphery: "We believe that the degree to which organisations combine the business and charity forms at their core, versus holding one at the periphery, follows a continuum rather than a dichotomy" (p. 425, *op. cit.*).

Thus non-profits may become increasingly commercial within new welfare markets as a result of isomorphic tendencies; and with increasing trends towards new public management, public sector organizations are becoming more socially entrepreneurial, operating in public sector constructed markets (and may become social enterprise through spinoffs from the public sector). Battilana and Dorado (2010) also note the spin-off process in the formation of commercial micro-finance organizations from NGOs.

To summarize: Hybrids may be bi-sectoral, combining the attributes of two sectors; they may be multi-sectoral, combining attributes of several sectors; and they may be intra-sectoral, combining attributes of different family members of a sector (such as CMAF). Hybridity may be at the core of the organization, or at its periphery – while a process of temporary/permanent transition takes place. Hybridity not only increases the general diversity of organizations within a field or sector, it also leads to a reappraisal of typologies of organizations within sectors; as well as the recognition of increasingly blurred boundaries, and difficulties in assessing boundary cases.

1.2. *Boundary cases*

Boundary cases are at the limits of definitional criteria, for example a criterion of being a formal organization might be operationalized through legal registration, but in some countries unregistered organizations may be large, and appear more formal than informal. Some definitional criteria appear more important for boundary cases than others; thus in the Johns Hopkins approach the private/self-governing criteria is particularly important as independence from government can be a current issue where "state services" are moving into the market, for example in the UK universities are formally charities, and hospitals are trusts. Whilst in the Pestoff (2005) diagrammatic definition of the third sector/social economy, there are only three boundary criteria: for-profit/non-profit, formal/informal, and public/private – here the limited distribution constraint of co-operatives would probably appear as a boundary case. And in the EMES nine dimensional definition of social enterprise (see below: section 3.1), although there are potentially nine types of boundary cases, some may be considered closer to the essence of social enterprise such as economic activity, and community benefit, rather than paid work, or collective entrepreneurship which might lose its relevance many years after the social enterprise is founded. But the ideal typical approach implies that boundary/dimensional criteria should not be applied in an exclusive manner, but overall judgments should also be placed in the context of performance against the other dimensions.

Mapping studies that survey populations of organizations using definitional criteria to identify a population of interest such as social enterprise, do not reveal the types of organizations that are boundary cases; they reveal difficulties of classification or difficulties in judging if an organization meets a particular criterion. Cluster analysis might reveal boundary cases, but an alternative approach to judging difficulties around each definitional criterion, is a typological analysis, where recognized types of organizations are examined in relation to the criteria. One typological approach may by based on different legal forms – thus for example new legislation for social enterprise in the UK that was passed in 2005; this legislation for the Community Interest Company was designed to fit the UK definition of social enterprise, but most social enterprises are formed using other legal forms (companies limited by guarantee) which are also used by a wide variety of third sector/ social economy organizations, including some which would not be considered social enterprises. An alternative typological approach is based on identifying self-labeling forms which, through collective association link with other similar organizations, for example Social Firms UK, has gathered together organizations which operate as work integration social enterprise; but the member social firms are quite varied, and not all would meet the UK definition of social enterprise. In another context many Eastern European countries have long established organizations for employing disabled people. Typically these are officially registered, thereby providing a sampling frame, and thus the initial discussion is whether to include this type in a mapping study of social enterprise. Further consideration of this type reveals that some of them have become for-profits, while others are non-profits, in other words only some would be likely to meet the criteria of social enterprise. A typological analysis is often needed to inform choice of sampling frames.

In general problems of mapping social enterprise require firstly a careful specification of the operational criteria for defining the phenomenon prior to surveying them; secondly a typological analysis focusing particularly on boundary case hybrids; and thirdly a choice of sampling frames which will encompass the population of interest.

2. Mapping the voluntary/non-profit sector in the UK

This section discusses issues of hybridity and boundary cases drawing on attempts to map the voluntary (non-profit) sector in the UK (Kendall and Knapp, 1993), within the framework of a research programme linked to the Johns Hopkins non-profit studies which utilized five criteria for defining non-profits. This involved consideration of definitional issues, population issues, boundary cases, and sampling frame issues. This led in the UK to broad and narrow definitions of the voluntary sector. Most countries that were part of the Johns Hopkins studies faced similar challenges, since there were different types of organizations on the boundaries of the non-profit sector and state/community/ informal/private sectors.

The term "loose and baggy monster" was coined by Kendall and Knapp (1995) to describe the diversity of organizations within the voluntary sector. Despite government regulation and a growing contract culture, this conceptualization is still very apt. And the growing trend towards hybridity, including through different forms of social enterprise, only serves to strengthen the validity of this view.

2.1. *Definitional issues*

Attempts to define and map the sector run headlong into this complex diversity; nonetheless considerable progress has been made in developing a statistical overview of the sector in many countries, and, going further to provide regular assessments of the state of the sector. The Johns Hopkins Comparative Non-profit Sector Project helped create a standard definition and the comparative database in many countries. Their approach provided a basis for defining a sector in each country, but the process involved a recognition of this diversity and the need to discuss boundary cases.

This mapping of the voluntary/non-profit sector in the UK was informed by the structural-operational definition of the non-profit sector of the Johns Hopkins Comparative Non-profit Sector Project which developed five key operational criteria that non-profit organizations share (Salamon and Anheier, 1992 and 1997) – they should be: Organized, Private, Non-profit-distributing, Self-governing, Voluntary.

It is noteworthy (Kendall and Knapp, 1996) that neither public benefit, nor social/charitable purposes are included in this set of criteria, because of the difficulties of reaching international agreement. And not only is this a problematic criterion in a comparative perspective, it is also longitudinally changing. Thus in the UK, there has been an evolving set of charitable purposes including (since 1601): repair of bridges, ports, havens, causeways, churches, sea-banks and highways. But after that, and for over 100 years there have been four types of charitable purpose: relief of poverty, education, advancement of religion, and trusts for other purposes beneficial to the community. And since the Charities Act 2011, there have been 13 categories of charitable activity.[2] This is highly

---

[2]   See: http://www.charitycommission.gov.uk/Charity_requirements_guidance/Charity_
essentials/Public_benefit/charitable_purposes.aspx).

relevant to the determination of criteria for defining social enterprise, particularly for helping to specify what is "social".

2.2. Population and sampling frame issues

Discussion of types of organizations to include within the non-profit sector, as well as discussion of boundary cases are central to the specification of the population (and sampling frame) to be considered in a mapping exercise.[3] Inevitably difficulties in applying some of the definitional criteria have led to discussion and debate over boundary cases. And it was an explicit part of the Johns Hopkins approach to identify and come to decisions about such boundary cases – thus for example universities in the UK are charities, but they may not be considered part of the third sector because they are highly regulated, and similarly for housing associations, where the UK "Government dictates the rules as regulator, supplies the capital grants, controls the Housing Benefit that makes up two thirds of rental income and fixes rent levels" (Purkis, 2010, p. 13). Reviewing the different types of organizations within a population helps identify boundary cases for further discussion. Part of the process of examining typologies may also be to consider populations in relation to international classificatory schemes; and the International classification of non-profit organizations (International classification of non-profit organization – ICNPO – developed by Salamon and Anheier, 1992) forms an important basis for differentiating between types of non-profit organizations, based on 12 different types of fields of activities (with sub-categories).

In the UK this led to *broad and narrow definitions* of the voluntary sector (Kendal and Knapp, 1996), partly because the ICNPO classificatory system tends to focus on service delivery, rather than mutual aid, advocacy and campaigning; and because the application of the 5 criteria is generally not clear, so boundaries are blurred. The broad definition was shaped by the structural-operational decision, while the narrow definition was an attempt to get closer to public understandings of "the voluntary sector" in the UK, since the 5 part structural-operational definition does not include a criterion related to charitable public benefit or altruism. So some organizations were excluded from the narrow definition because there was a lack of recognition that they were part of the voluntary sector ("would probably not feature in most people's understandings of the voluntary sector in the UK" (Kendall and Knapp, 1996, p. 21).

---

[3] See in this book chapter authored by Bouchard, Cruz Filho and St-Denis.

284

2.3. Boundary cases

These points are illustrated with respect to a number of boundary cases in the UK, including: recreational organizations, universities, and schools. The following types of organizations were excluded from the narrow definition, for following reasons:

- Recreation (sports clubs, recreation and social clubs) – because they seem to lack an "altruistic core," which Kendall and Knapp (1996) argue is part of a common understanding of what is the voluntary sector.

- Primary and secondary education – part of this sector consists of private schools which have charitable status, but require high fee payments, so are rather exclusive, rather than being altruistic; the other part of this sector, although governed by citizens, are totally dependent for resources from the state, so have been considered as part of the state sector.

- Higher education – although many such institutions are registered charities, including universities, until recently most of their funding has been from the state, thus they have not been considered sufficiently independent, although this has been changing due to large increases in student fees.

- Business, professions and trades unions (excluded because they are not generally charitable, and lack a "charitable core").

But they were included in the broad definition of the voluntary sector; and the application of the criteria was difficult, thus for example: housing associations were considered part of the narrow voluntary sector, and universities, the broad voluntary sector, despite both being highly regulated by government. And this may be an indication that boundary cases and hybridity apply to all sectors, and a wide range of fields – including public services and business. The following diagram summarizes the broad types of organizations within the broad definition, and shows areas of the boundary cases, relating to the criteria: formal, independent, non-profit distributing.

And note that certain categories (political organizations and sacramental organizations) were excluded from both the broad and the narrow voluntary sector population for statistical reasons (and there may be historical reasons for exclusions cf. Bouchard *et al.*, 2011).

Figure 1. *Voluntary sector diamond*

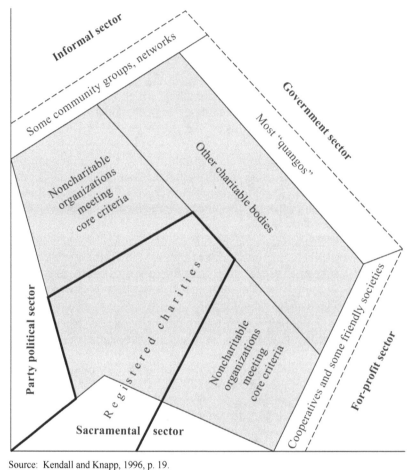

Source: Kendall and Knapp, 1996, p. 19.

Thus for example, on the boundary between the voluntary sector and the government/state are group of organizations known as quangos – quasi-autonomous non-governmental organizations (see Box 1 below); and these state hybrids bear considerable resemblance to the highly regulated voluntary sector hybrids – the main difference being governance, where the board is typically nominated by government.

Box 1. The growth of quangos

The growth of quangos

The first report of the Committee on Standards in Public Life (1995) was chaired by Lord Nolan, and had one of its four chapters dedicated to quangos (quasi-autonomous non-governmental organizations), hybrid organizations on the boundary of the public sector. These non-departmental public bodies (NDPBs) including National Health Service bodies carry out governmental functions at arm's length from the minister. The main concerns were: appointments – whether they were being unduly influenced by political considerations, and whether they were sufficiently open; and secondly propriety (openness, independent monitoring, and accountability). And while the was a decline in the number of in NDPBs from 2167 in 1979 to 1003 and 45 in 1994, they had almost doubled their public expenditure (in 1994 prices) from £8.5 billion to £15.08 billion. (Note that major NHS bodies totaled 770, with £25 billion expenditure). Some of the major NDPBs include higher education funding Council, Housing Corporation legal aid board, Scottish enterprise, British Council, Housing for Wales, English Heritage. The Nolan report also notes that there are wider group of quangos which are "independent self-governing, but which spend public money and perform public functions," these included training and enterprise councils, local enterprise companies housing associations, further education corporations and ports or grant maintained schools. The report notes that membership is usually regulated by statute or in contracts with government.

In a longitudinal perspective, housing associations have also proved a changing boundary case; they are highly regulated third sector organization, similar to hybrids in the adjoining public sector where reforms to quangos (quasi non-governmental organizations), have had an influence on whether housing associations have been classified as a boundary case within the third sector, or the public sector. In a different longitudinal perspective, the typological and boundary discussion process can lead to reclassifications of boundary cases (for example in Italy non-profit researchers reclassified social co-operatives as non-profits).

Quangos arise from a recent hybridization process, but remain outside the boundary of the voluntary sector. New hybrid legal forms such as that for social co-operatives in Italy combine some characteristics of non-profits with those of co-operatives (in the Johns Hopkins research programme this has led to a reclassification, and repositioning of that form into the non-profit field). But boundary cases arising from processes of hybridization do not necessarily lead to new legislation – since existing law can be very flexible. For example in Sweden, the researchers noted that the new hybrid form of social co-operative typically used their "ideell" association legal form, which was commonly used by non-profits, and thus proposed "to include in the definition of a Swedish non-profit sector these "neo-co-operatives" (such as for childcare) along with the different forms of business associations, but to exclude from it the tradition al co-operative movement organizations" (Lundström and Wijkström, 1995, p. 13).

2.4. Sampling frame and mapping issues

A major consideration in establishing sampling frames is the legal structures used by a population of organizations, since registration systems are often set up for different legal forms. In certain countries there may be an identity between the population of the third sector social economy or its component parts (CMAF), and the legal forms used by the population of organizations. But in other countries, such as the UK, there are a few legal structures used for different kinds of organizations, but they do not easily map onto the non-profit/voluntary sector, or the co-operative sector, and certainly not the social enterprise sector – as defined in the UK, virtually any legal form could be used.

Nonetheless legal structures are an important consideration in the identification of sampling frames.[4] In the UK voluntary sector the following types of legal structures are used: company limited by guarantee (a democratic, incorporated association) is the most common form; charitable trusts (charities), foundations (grant giving), unincorporated associations (i.e. without limited liability), and other forms such as friendly societies, industrial and provident societies (particularly those for the benefit the community), also housing associations have a separate structure.

There are databases (kept by the registration bodies) for each of these legal structures, but several of the legal structures are not exclusively for voluntary organizations – for example companies limited by guarantee is the most popular legal structure for both voluntary organizations and co-operatives; secondly not all charitable bodies are required to register (exempted, and excepted charities); and so on.

An alternative approach is based on employment statistics. In some countries governments collect data on employment in different industries, and separate out the ownership sectors; this provided the basis for statistical data in France, Germany and the United States. However such data was not available in the UK.

---

4   See in this book chapter authored by Archambault.

In the mapping exercise of non-profits in the 1990s, the sampling frames for surveys were based on registers of different relevant legal forms; and they had the deficiencies noted above. Subsequently the sampling frames have been improved, so that mapping the sector is now more straightforward, because the third sector gained considerable political support under New Labor (the Blair led Labor Party government from 1997-2010). However Kendall and Knapp in the early 1990s, faced with these difficulties, adopted a *triangulated mapping strategy* (described below).

In the absence of good comprehensive sampling frames, mapping requires considerable creativity in developing appropriate data gathering strategies. Kendall and Knapp (1996) developed a modular approach which they describe as GUSTO: because it is based on the use of **government** statistics, voluntary sector **umbrella** body tabulations, **secondary** analysis of these data, **territorial** surveys (to cross validate or inform assumptions about the full picture), and **original** organizational surveys.

It is also important to consider the comparability of the different databases, for example is there a registration threshold in terms of size or trading income, et cetera.

3. Mapping social enterprise

This Section continues by discussing various attempts by the UK's Third Sector Research Centre to map the third sector, and social enterprise, which involved addressing some of the challenges of boundary cases and hybrids. In the case of the third sector, Mohan (2011, p. 4) notes "… in terms of entities with at least some recognizable degree of organization, numbers of third sector organizations might vary by a factor of as much as nine". And a similar picture for social enterprise is reported by Lyon and Sepulveda (2009), but here in relation to government surveys. In the former case this is largely due to "beneath the radar" organizations that are not formally registered; while in the latter case there is considerable definitional uncertainty about social enterprise using for-profit structures. A critical perspective on the UK approach is developed, by drawing both on the approach of the Johns Hopkins non-profit studies, as well as using the EMES nine part dimensions for defining ideal typical social enterprise.

3.1. Definitional issues

The UK definition of social enterprise is: "a business with primary social/environmental objectives, whose surpluses are principally reinvested for that purpose in the business or community rather than mainly being paid to shareholders and owners." This definition is clearly business oriented, unsurprisingly because it derives from the time when the social enterprise unit was based in the Department of Trade and Industry (DTI, 2002). As we shall see this business oriented definition creates problems for certain third sector organizations to identify themselves as social enterprise.

The definition is operationalized as follows:

- the organization is trading, and generates a certain percentage of its income (25/50%) from trading of goods and services in private or public markets;
- primary purpose is to provide social/environmental goals, rather than purely for profit;
- the organization principally reinvests profits/surplus into the organization or community to further social/environmental goals (typically operationalized as >50%);
- and some surveys have included a self-identification criterion: asking if the organization sees itself as a business with primary social/ environmental objectives, etc.

This operationalized definition is now critically examined then compared to the EMES approach (Defourny and Nyssens, 2008).[5]

The first criterion – trading – some non-profits trade, other don't; thus some but not all social enterprises would be non-profits.

The second criterion – social/environmental goal – is certainly the most difficult to operationalize; indeed as Kendall and Knapp (1996, p. 20) argued whereas one might expect "…'altruistic' or operate for public benefit' to be included as part of the voluntary sector (definition)… This criterion was *not* employed for the purposes of cross national comparisons since it was not possible to agree on how it might be applied in practice internationally." However within the next stage of mapping – defining the population of interest – standard industrial type classifications were used to specify 'social' activities, based on an international classification of non-profit organizations (ICNPO).

The third criterion – on reinvestment – allows 49% profit distribution, which is much more liberal than the non-profit distributing criterion where no profit is distributed. Profit distribution might be expected, since part of the hybrid character of social enterprise would be to have some business-like elements; and for example co-operatives which have a limited distribution constraint, are generally considered part of the social

---

5   See EMES' website: http://www.emes.net/about-us/focus-areas/social-enterprise/.

enterprise sector. But this is much more than the average for listed compa-nies on the US Standard and Poor's (S&P), stock market index where the "Dividend Payout Ratio" is around 29% profit distribution (JP Morgan, 2012), but it has had an average of 42% over the last 23 years. And there is some inconsistency between how the limits on profit distribution are legally specified in the UK's Community Interest Company (CIC), which is the legal form recently developed (2006) specifically for social enter-prises, and this reinvestment criterion. The CIC profit distribution is at a lower level than 49% – it is currently capped at 35%, while dividend distribution (based on a proportion of capital invested) is capped as 20% (having been raised from 5% above base rate).

The fourth criterion – self-identification as a social enterprise – is interesting from a policy perspective, but it would be most unusual to include it as part of the scientific basis for defining the population of the field; particularly as social enterprise is an emerging field compris-ing new social enterprises, and existing organizations reconfiguring and restructuring due to internal and contextual factors – as evidenced in NCVO Almanac (2009) which for a number of years has specified the amount of social entrepreneurial activity carried out in the voluntary sec-tor. And in the UK Annual Small Business Survey (ASBS), which sur-veyed organizations using a diverse range of legal structures including those legally registered as sole proprietors with no employees (just under 50% of businesses surveyed), and those legally registered as partnerships (over 15%), a self-identification question asking whether they considered themselves to be small businesses was not used to define the population small businesses.

However although not specified in the UK definitional criteria, there are three other important criteria that appear relevant to defining the so-cial enterprise sector: *formally organized, private, independent or self-governing*. Even if one of the criteria (*organized*) is implied in the choice of a sampling frame of formally registered organizations,[6] *private* and *independent/self-governing* is a much more problematic criterion to ap-ply, with boundary cases regarding independence from government – and this may lead two important choices in relation to a sampling frame using broad versus narrow definitions of the voluntary sector.

---

[6]   Note that although this criterion (organized/formal) is important for policy makers, and facilitates research since sampling frames for informal organizations generally don't exist, there is considerable interest in this "below the radar" area of associa-tive activity (see for example McCabe, Phillimore, and Mayblin, 2010). And the ILO has developed an approach for measuring informal activity: http://www.ilo.org/stat/ Publications/WCMS_182300/lang--en/index.htm.

Before turning to consider a more developed set of criteria, that of EMES, it is worth noting a European definition linked to the Social Business Initiative (2011), which embraces many of EMES dimensions:

> A social enterprise is an operator in the social economy **whose main objective is to have a social impact rather than make a profit** for their owners or shareholders. It **operates by providing goods and services for the market** in an **entrepreneurial** and **innovative** fashion and **uses its profits primarily to achieve social objectives**. It is **managed in an open and responsible manner** and, in particular, involves employees, consumers and stakeholders affected by its commercial activities.[7]

– EMES Comparison:

It is useful to test the operationalization of the UK definition against the EMES definition, which is an internationally recognized ideal typical approach based on extensive in internationally comparative empirical studies of social enterprise.

This EMES definition with its nine dimensions (Defourny and Nyssens, 2008) was developed in an ideal typical perspective as a basis for examining specific social enterprise, and types of social enterprise. The outcome of such an examination is not a strict boundary between the population of social enterprise and non-social enterprise, but a very much more hybridized spectrum, in a world of increasing hybridity. Thus a core of social enterprise would be closely aligned with all the dimensions, while others would have a more peripheral fit – not fitting certain dimensions, or being boundary cases. The EMES approach was not developed to be easily operationalizable in surveys of the social enterprise field. Nonetheless these dimensions represent a more complete richness of the characteristics of social enterprise, and serve as a useful basis for critical comparison and development of operational criteria. The following explores how the EMES dimensions may be used in a more operational manner.

[Note the operationalized dimensions proposed in this chapter has been developed by the author and is not part of the EMES approach; see also a very comprehensive approach adopted for mapping the social economy in Quebec (Bouchard *et al.*, 2011)].

We now critically examine the UK approach, by developing and using operationalized dimensions (OD) of each dimension of the EMES approach.

---

[7] Communication from the Commission to the European Parliament, the Council, the European Economic and Social Committee and the Committee of the Regions: *Social Business Initiative – Creating a favorable climate for social enterprises, key stakeholders in the social economy and innovation*; COM (2011) 682; Brussels, 25.10.2011.

The first EMES dimension requires assessing whether the organization is conducting "a continuous activity producing goods and/or selling services"; while the second EMES dimension assesses whether the organization engages in "a significant level of economic risk; both these can be operationalized into 'percentage of income from trading'; the UK definition emphasizes trading which implies economic activity and economic risk; but the EMES framework recognizes economic risk associated with other resources (such as competing for grants/subsidies/donations), whilst risk in some markets may be low (due to entry barriers, captive markets, etc.), hence the proposed low threshold OD of 25%.

The third EMES dimension specifies "a minimum number of paid workers"; it is proposed that an OD with a low threshold of one full-time equivalent (FTE) employee could be adopted. This is not included in the UK definition, but in the UK survey data one or more paid workers has been applied as a criterion.

The fourth EMES dimension – "an explicit aim to benefit the community" – is more specific than the UK definition, but interestingly this is how the social purpose is specified in regulations for its community interest company. The EMES dimension specifies that it should be a primary aim to benefit the community; operationalizing this raises potential difficulties with subjectivity of survey responses. The UK surveys have a broader dimension – social/environmental purpose – and this is operationalized through statements by senior managers in response to questionnaires; both approaches are subject to the challenge that they lack sufficient clarity or exemplars to ensure consistent responses; a more rigorous alternative would be to request a statement of activity, then use a coding framework to classify responses.

The fifth EMES dimension refers to a collective civil society entrepreneurial process – "an initiative launched by a group of citizens or civil society organizations"; operationalizing this would require a classification of different entrepreneurial dynamics: organizational, group of citizens, individual; entrepreneurship could be further differentiated to recognize both de novo start-ups, and intrapreneurship from established social enterprise. However this does not appear to be considered relevant in the UK approach; and it may appear less relevant than governance, the longer an organization has been in existence.

The sixth EMES dimension – "a high degree of autonomy" – accords with the non-profit approach of being private/independent and self-governing, and has already been discussed above.

The seventh EMES dimension – "a decision-making power not based on capital ownership," and the eighth – "a participatory nature, which involves parties affected by the activity" i.e. stakeholders participation;

there are no ODs included in the UK approach that are similar to this dimension, but social enterprise that use third sector legal forms (such as Company Limited by Guarantee (CLG)) would generally fulfill these criteria. But for other sampling frames a typology of stakeholders and their level of participation (including in governance) would need to be developed and used.

Finally the ninth EMES dimension – "a limited profit distribution" raises some challenges for the UK reinvestment criterion which would allow 49% profit distribution, since this would be above conventional understandings about 'limited profit distribution'; and as noted above this differs from the UK community interest company (CIC) which has a lower distribution limit (35%).

Finally, in general terms the EMES approach emphasizes governance dimensions more strongly than the UK approach, it is less business oriented and more open to a hybridized resource base, and as result of the latter, it is more open to recognizing informal structures particularly those arising from the solidarity economy.

Approaches to defining social enterprise also draw on a consideration of the different activities that social enterprise carry out, but developing a typology similar to that for non-profits (ICNPO) is more difficult to apply, since social enterprise operate in a much wider range of sectors. Bouchard *et al.* (2008) also argue the ICNPO classification system is not homogeneous, but combines both social activities and social mission.

– The self-identification criterion:

A self-identification question has been used in surveys of social enterprises in the UK. The self-identification question is: "social enterprises are businesses with primarily social objectives whose surpluses are principally invested for that purpose in the business or community, rather than being driven by the need to maximize profit for shareholders and owners. Does this describe your organization or not?" (Q38 in National survey of third sector organizations, 2010).

There are two subcategories of the population of respondents for which this is a challenging question: social enterprises which do not meet the three definitional criteria, but which see themselves as social enterprises; and the subcategory that fulfills the three definitional criteria, but doesn't see itself as a social enterprise (for various reasons).

There are 2 paths to the development of social enterprise: new social enterprise, and the reconfiguration of existing structures like non-profits to be more socially entrepreneurial. Thus there could be a number of reasons why organizations don't self-identify as social enterprise. It may be the case that the issue is around the definition of social aims (Lyon and

Sepulveda, 2009). However it could also be that the formulation of the self-identification question requires an organization to identify itself as a business, and clearly many organizations that match the other definitional criteria of social enterprise may be charities conducting a substantial amount of social entrepreneurial activity, but for obvious reasons would not self-identify as a business – since they are registered as charities.

Some researchers (e.g. Lyon, Teasdale, and Baldock, 2010, p. 9) argue that the self-identification question should be used as a basis for defining the population of social enterprise in the UK. However this seems problematic for the above reasons. An alternative view would be that either the question needs to be reformulated, or the definition of social enterprise needs to be adapted for example to be more inclusive of third sector organizations becoming more entrepreneurial, for example: "social enterprises are *organizations* with trading income, whose objectives are primarily social, and whose surpluses are principally invested for that purpose in the business or community, rather than being driven by the need to maximize profit for shareholders and owners" (…rather than insisting they should categorize themselves as a "business").

Alternatively the question could be elaborated to be more inclusive of third sector and specify different types of social enterprise including businesses with a social goal, and voluntary and community social enterprise, including this category of charitable hybrid – through terminology such as "third sector social enterprise". Or a set of questions could be presented in order to explore the different attributes of "belonging" to a sector.

3.2. Defining the population of social enterprise

As noted in Section 1, hybrids may be:

- bi-sectoral, combining the attributes of two sectors;
- multi-sectoral, combining attributes of several sectors,
- intra-sectoral, combining attributes of different family members of a sector (such as CMAF)

One approach to defining the population of social enterprise is to argue that a social enterprise sector is within the social economy, and this is the position of the EMES network which initially took an intra-sectoral perspective with social enterprise combining attributes of nonprofits and cooperatives (and more recently has taken a multisectoral approach – Evers and Laville, 2004). In this perspective, defining the population requires some reflection on excluding those members of the social economy which are not social enterprise; this would include identifying those organizations which are not trading or not sufficiently in the market (according to the definitional criteria specifying the proportion of income that should be traded); it might also exclude organizations that are not considered sufficiently social – there have been frequent debates over the extent to which large and long-established co-operatives and mutuals which have become isomorphic with conventional business should be considered social enterprise, and which of their social enterprise characteristics have changed/degenerated.

Figure 2. *The fields of the social economy/social enterprises*

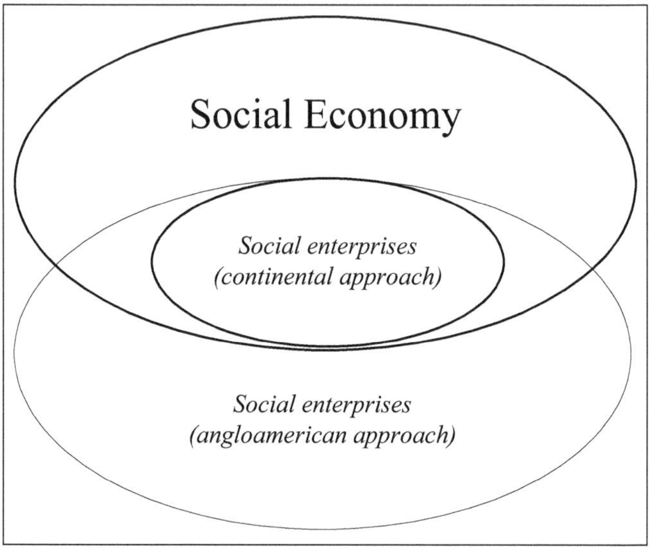

Source: Monzón and Chaves, 2012.

However other views would lead to a wider range of sampling frames: some (especially from the European social economy) would argue on the basis of social goals, that social enterprises are more narrowly defined, within the Solidarity Economy with its emphasis on social values. But in contrast, Anglo-Saxon approaches (see Table 1 below) and approaches from outside mainland Europe would adopt a bi-sectoral approach with social enterprise combining elements of non-profit charities and business. In this perspective, the population includes organizations both from within the social economy and from outside that sector, to include socially oriented business structures. And since hybridity may be at the core of the organization, or at its periphery – while a process of temporary/permanent transition takes place – the boundaries are rather fuzzy.

In the UK, the views of key stakeholders on the population of social enterprise are different, if not contested. And these different perspectives on the field have led to very different survey data, and revealed different boundary cases: in contrast to the European continental perspective (EMES), where for-profit social enterprise are boundary cases, in the UK surveys (ASBS in Table 1 below), an Anglo-Saxon perspective has resulted in the identification of a very large proportion of for-profit social enterprise, including those with no employees – although these judged an excluded boundary case, because of no paid employment; in contrast social enterprise using third sector legal structures (e.g. Company Limited by Guarantee), which might be considered a prominent type, only formed a small proportion of the total, largely due to the small business sampling frame selected. And from an entirely different perspective, in Eastern European countries, quasi-public social enterprises might be considered boundary cases due to a gradual restructuring of state organizations.

An alternative approach is proposed by Mohammed Yunus (2009) who specifies two categories of social business – one of which is a for-profit that re-invest all their profits (no dividend distribution) and in the other type – co-operatives owned by the poor – he emphasizes the disadvantaged as central stakeholders.

3.3. Sampling frame issues

In the UK social enterprises may use a wide range of legal structures, but only one of these, the Community Interest Company, is specifically for social enterprise. The possible legal forms are: Company Limited by Shares (CLS), Public Limited Company (PLC), Partnerships, Industrial and Provident Society, Company Limited by Guarantee (CLG), Friendly Society, Community Interest Company (CIC).

Recent surveys in the UK have adopted two contrasting approaches: either adopting a sampling frame of the third sector, or a business sampling frame. This has led to 2 rather distinct population estimates of social enterprise, based on two broad types of social enterprise – third sector social enterprise, and private-sector social enterprise.

3.4. Mapping strategies and outcomes

There have been at least five widely different survey-based estimates of population of social enterprise since 2005 in the UK – as summarized by Lyon, Teasdale, and Baldock (2010). While it might be possible to develop broad and narrow perspectives on this diversity, the first step would be to recognize typological differences between third sector and private-sector social enterprise.

The UK definitional criteria have been applied differently, but it is possible to reconcile and harmonize the criteria for the different survey estimates (table below adapted from Lyon, Teasdale, and Baldock (2010)).

Table 1. *UK Surveys of social enterprise*[8]

Data source	Description	Sample details	No of orgs (000s)	T/O (£bns)	Empl (000s)	Comments
IFF 2005	> 25% income from trading and self defining	Only CLG and IPS	15	18	475	
ASBS narrow (2005-7)	SE with employees	Dominated by private, under-representing TS	70	15.5	248	Only 8,000 in a third sector legal form
ASBS wide (2005-7)	All enterprises meeting SE tests	As above	234	23.6	410	Only 10,000 in TS legal form
NSTSO narrow	>50% income from trading and self defining	Third sector only	16	8.5	227	
NSTSO wide	>50% income from trading but not self defining	As above	21	10.7	272	
NCVO (2009)	SE activity	All civil society		77		
Delta (2010)	Businesses wanting to make a difference	Private, less than 2 years old with >£200K income	232	97		Not self-defining

Only two of the above surveys focused on the official small business database (ASBS), and the third sector organizations within the surveys can be relatively easily identified. Thus using the three main operational criteria (trading income > 50%,[9] reinvesting majority of profit, and social purpose) the population has risen from 15,000 in 2005 to 21,344 in 2010 – this represents the population of third sector social enterprise (and excludes the self-identification test for reasons argued above). There may be broad population estimates substantially larger than this but these probably include some of the broad voluntary sector organizations noted earlier (such as universities?).

---

[8] Key: IFF: IFF Research (2005) *A Survey of Social Enterprises Across the UK*, London: Small Business Service; NSTSO: Ipsos MORI (2009), *National Survey of Third Sector Organizations: Technical report*, UK Data Archive Study Number 6381: *National Survey of Third Sector Organizations, 2008-2009*; Delta: Delta Economics (2010) *Hidden Social Enterprises*. London: Delta Economics/IFF Research. http://www.deltaeconomics.com/Uploads/Misc/HSE2010final4.pdf.

[9] Note the most recent UK survey (BMG, 2013) rather confusingly increases the trading income criterion to 75%.

The other population surveyed is the business sampling frame, and if the small number of third sector organizations in the sample are excluded, and the results normalized to the three criteria – in particular adjusted to reflect 50% trading income – this reduces the headline figure of 62,000 social enterprise to 59,520 social enterprise; and these reflect the private sector social enterprise.

This gives a combined total of 80,866 social enterprises in the UK (approximate date 2010). As Lyon, Teasdale, Baldock (2010) note, government policy seems to be oriented to third sector capacity building and social enterprise development, while the headline figure of 62,000 social enterprises refers to private-sector social enterprise. These authors take an even more critical stance in their 2013 paper (Teasdale, Lyon, Baldock), and question the extent of growth in the sector – indeed changes to the operational criteria pose considerable challenges to longitudinal analysis of performance.

The figure of 59,520 private-sector social enterprises requires further study to examine social purpose, and whether the reinvestment criterion of 50% is appropriate – given points made above about 35% profit distribution for CICs.

3.5. Summary

To conclude with comments on the UK attempts to map social enterprise, the following measures would improve the operationalization of the UK definition:

- Recognition that the following criteria are important: *organized/ formal*, *private*, *self-governing*; and thus a more complete set of operational criteria would need to include them;

- Develop consistent criterion for 'percentage income from trading'; the UK criterion has fluctuated from 25% to 50% to 75% most recently. It is also important to recognize that economic risk doesn't only arise from trading; the EMES framework recognizes economic risk associated with other resources; this might justify use of a low threshold of 25%.

- Make explicit that the threshold of one full-time equivalent (FTE) employee has been employed as a criterion in reporting UK statistics on the sector.

- Improve the social purpose criterion which suffers from a lack of sufficient clarity or exemplars to ensure consistent responses (and coherent links with CIC criterion of "community benefit").

- Establish consistency of criteria between different types social enterprise, particularly where for-profit sampling frames are used; particularly regarding governance; a typology of stakeholders and their level of participation (including in governance) could be developed and used.

- UK reinvestment criterion would allow 49% profit distribution, but this would be above conventional understandings about 'limited profit distribution'; and far higher even than Public Limited Company (which are quoted on the stock exchange) profit distribution norms; and this differs from the UK community interest company (CIC) which has a lower distribution limit (35%).

Furthermore the definition of social enterprise could be adapted as indicated above to be more inclusive of established third sector organizations developing substantial social entrepreneurial activities, which could be considered as social enterprise hybrids emerging as a result reconfiguration of the field; this would facilitate self-identification (although self-identification doesn't seem a rigorous criterion for defining the scope of the sector); this could be reformulated as:

> Social enterprises are *organizations* with trading income, whose objectives are primarily social, and whose surpluses are principally invested for that purpose in the business or community, rather than being driven by the need to maximize profit for shareholders and owners.

Conclusions: Hybridity, typologies and boundaries

The chapter now moves to its main conclusions: Hybrids may emerge in two main ways – either through the development of new organizations, or through the transformation of existing organizations. In examining these different pathways, it is useful to recognize that hybridity may be at the core of the organization, or at its periphery, in which case a process of temporary or permanent transition may take place.

Hybrids may be multi-sectoral, combining attributes of several sectors, alternatively they may be intra-sectoral, combining attributes of different family members of a sector (such as CMAF). Social economy researchers have argued these patterns explain hybridization within the social economy, and the emergence of social enterprise as an intra-sectoral hybrid of non-profits and co-operatives. Other researchers have identified a bi-sectoral hybrid combining the attributes of non-profits and business – and these are problematic to research, as the hybrid emerges from different populations.

There is an increasing trend towards hybridity, but isomorphic tendencies may exacerbate the difficulties of mapping social enterprise, as they influence the stability of hybrids. But institutionalizing processes (e.g. new legislation/policy) will consolidate patterns of hybridity into new typologies in the social economy, as well as on boundaries with other sectors.

It is argued that

there are two dynamics in the hybridisation process of existing organisations: *contextual* and *internal*. Internal factors include a strong motivation towards a social mission, and a desire to be more entrepreneurial and gain access to market resources; and other types of factors such as: oligarchic tendencies (the power of managers and staff dominance), decline in member participation, and the growth of 'commoditised' member relations; the tension between business and mission related activities (and their separation through for example trading subsidiaries, or board roles) (Spear, 2010, p. 4).

Contextual hybridization processes include: the institutionalization of new legal forms (either as part of a policy initiative to promote social enterprise as in UK), or as a result of civil society led institutional innovations which have been replicated sufficiently to become a new hybrid, and subsequently gain recognition in law (e.g. the Italian social co-operative legislation in 1991). The other major contextual hybridization process is the increasing marketization of areas of activity – quasi-markets for public services, individual personal budgets, markets for donations, etc.

This chapter has shown that when mapping a sector, such as social enterprise, there are challenges associated with definitional issues, population and sampling frame issues, and mapping strategies. Depending on choices in relation to these issues, different boundary cases and hybrids will come into focus. A typological analysis will help reveal different kinds of hybrids; the challenges associated with those hybrids that have become institutionalized within the social economy (e.g. social co-operatives combining elements of co-operatives and non-profits) are more straightforward than emerging hybrids that have a lesser degree of institutionalization such as membership of self-labeling groupings, but which may be rather heterogeneous. Thus civil society actors and policy makers both contribute to the institutionalization process. But researchers also help identify hybridization trends, for example as Buckingham (2011) notes, nonprofit organizations engaged in public service contracting may be subdivided into comfortable, compliant, and cautious contractors; this seems to indicate a potential further subdividing of these non-profit social enterprise hybrids.

Definitional issues are typically problematic for researchers, since establishing consensus on an emerging field requires discussions and negotiations not only with other researchers, but with policy makers and civil society actors. As noted in the discussion of UK social enterprise mappings (Teasdale, Lyon, Baldock, 2013), policy makers can "muddy the waters" by inconsistencies and incoherencies in survey approaches.

It also has to be recognized that the more problematic boundary cases are those across the major sectors – public/private/social economy;

not least because these may be contested; thus for example in the UK, for-profit social enterprise which comprise the main sampling frame are hardly discussed at the policy level, while third sector social enterprise are the major theme in policy discourse; many people are not aware of this disconnect between survey data, and discourse; and they would have strong reservations about the concept of for-profit social enterprise. However this approach of a typological and boundary discussion process seems a fruitful path for defining increasingly hybridized sectors.

References

Battilana, J., Dorado, S., "Building sustainable hybrid organizations: The case of commercial microfinance organizations," *Academy of Management Journal*, No. 53, 2010, pp. 1419-1440.

Battilana, J. and Lee, M., "Advancing Research on Hybrid Organizing Insights from the Study of Social Enterprises," *The Academy of Management Annals*, Vol. 8, No. 1, 2014, pp. 397-441.

Bode, I., Evers, A., Schulz, A., "Work Integration Social Enterprise in Europe: Can Hybridization Be Sustainable?," in Nyssens, M. (ed.), *Social enterprise: At the Crossroads of Market, Public Policies and Civil Society*, London, Routledge, 2006, pp. 237-258.

BMG Research, *Small Business Survey 2012: SME Employers*, UK Government, Department for Business, Innovation, and Skills (BIS), 2013.

Borzaga, C., and Defourny, J., *The Emergence of Social Enterprise*, London, Routledge, 2001.

Bouchard, M. J., Cruz Filho, P., St-Denis, M., *Cadre conceptuel pour définir la population statistique de l'économie sociale au Québec*, Cahier de la Chaire de recherche du Canada en Economie sociale, Collection Recherche No. R-2011-02, 2011.

Brah, A., Coombes, A. E. (eds.), *Hybridity and its Discontents: Politics, Science, Culture*, London, Routledge, 2000.

Brandsen, T., van de Donk, W., Putters, K., "Griffins or Chameleons? Hybridity as a Permanent and Inevitable Characteristic of the Third Sector," *International Journal of Public Administration*, Vol. 28, No. 9-10, 2005, pp. 749-65.

Buckingham, H., "Capturing Diversity: A Typology of Third Sector Organisations' Responses to Contracting Based on Empirical Evidence from Homelessness Services," *Journal of Social Policy*, Vol. 41, No. 3, 2011, pp. 569-589.

Committee on Standards in Public Life, *Standards in Public Life: First Report of the Committee on Standards in Public Life*, London, UK Government, Her Majesty's Stationary Office, 1995.

Cornforth, C. J., Spear R., "Hybrids and Governance: Social Enterprise," in Billis, D. (ed.), *Hybrid Organizations and the Third Sector: Challenges of Practice, Policy and Theory*, London, Palgrave MacMillan, 2010.

Côté, D. (ed.), *Les holdings coopératifs. Evolution ou transformation définitive?* Brussels, De Boeck Université, 2000.

Defourny, J., and Nyssens, M., "Social Enterprise in Europe: Recent Trends and Developments," *Social Enterprise Journal*, 2008, Vol. 4, No. 3, pp. 202-228.

Defourny, J., and Nyssens, M., "The EMES Approach of Social Enterprise in a Comparative Perspective," *EMES European Research Network*, 2012, WP No. 12/03.

Evers, A., "Mixed Welfare Systems and Hybrid Organizations: Changes in the Governance and Provision of Social Services," *International Journal of Public Administration*, 2005, Vol. 28, No. 9, pp. 737-48.

Evers, A., and Laville, J.-L., "Defining the Third Sector in Europe," in A. Evers and Laville, J.-L. (eds.), *The Third Sector in Europe*, Cheltenham, Edward Elgar, 2004, pp. 11-37.

Lounsbury, M., and Boxenbaum, E., *Institutional Logics in Action*, Bingley, Emerald Books, 2013.

Lyon, F., and Sepulveda, L., "Mapping Social Enterprises: Past Approaches, Challenges and Future Directions," *Social Enterprise Journal*, Vol. 5, No. 1, 2009, pp. 83-94.

Lyon, F., Teasdale, S. and Baldock, R., *Approaches to Measuring the Scale of the Social Enterprise Sector in the UK*, Third Sector Research Centre Working Paper No. 43, 2010.

Lundström, T. and Wijkström, F., *Defining the Non-profit Sector: Sweden*, Working Papers of the Johns Hopkins Comparative Non-profit Sector Project, WP16, 1995.

McCabe, A., Phillimore, J. Mayblin, L., "'Below the Radar' Activities and Organisations in the Third Sector: A Summary Review of the Literature," *Third Sector Research Centre Briefing Paper* 2010, No. 29.

Mohan, J., *Mapping the Big Society: Perspectives From the Third Sector Research Centre*, Third Sector Research Centre Working Paper, No. 62, 2011.

Monzón, J. L., and Chaves, R., *The Social Economy in the European Union. Presentation to European Economic and Social Committee*, Brussels, 3 October 2012. http://www.eesc.europa.eu/resources/docs/presentation-by-dr-mr-monzon.pdf.

Morgan, J. P., *Distribution Policy: Dividend and Share Repurchase Facts and Trends*, Corporate Finance Advisory, 2012. https://www.jpmorgan.com/cm/BlobServer/JPMorgan_CorporateFinanceAdvisory_2012DistributionPolicy.pdf?blobkey=id&blobwhere=1320577225001&blobheader=application%2Fpdf&blobheadername1=Cache-Control&blobheadervalue1=private&blobcol=urldata&blobtable=MungoBlobs.

NCVO, *UK Civil Society Almanac*, London, National Council of Voluntary Organisations, 2009.

Ostrom, E., *Self-Governed Common-Pool Resource Institutions*, Cambridge, Cambridge University Press, New York, 1990.

Pestoff, V., *Beyond the Market and State. Civil Democracy and Social Enterprises in a Welfare Society*, Aldershot, UK and Brookfield, NJ: Ashgate, 1998 & 2005.

Purkis, A., *Housing Associations in England and the Future of Voluntary Organisations*, London, The Baring Foundation, 2010.

Spear, R., *Co-operative Hybrids.* Keynote presentation at "Co-operatives' contributions to a plural economy," Conference of Research Committee of the International Co-operative Alliance with the CRESS Rhone-Alpes and the University Lyon 2; 2-4 September 2010.

Spear, R., "Hybridité des coopératives," in Blanc, J. and Colongo, D. (eds.), *Les contributions des coopératives à une économie plurielle*, Paris, L'Harmattan, Les cahiers de l'économie sociale, Entreprendre autrement, 2012.

Teasdale, S., Lyon, F., Baldock, R., "Playing with Numbers: A Methodological Critique of the Social Enterprise Growth Myth," *Journal of Social Entrepreneurship*, 2013, Vol. 4, No. 2, pp. 113-131.

Young, D. R., *If Not For Profit, For What? A Behavioral Theory of the Nonprofit Sector Based on Entrepreneurship*, Lexington, Mass., Lexington Books, 1983.

Yunus, M., *Creating a World without Poverty: Social Business and the Future of Capitalism*, New York, Public Affairs, 2009.

Conclusion

A Research Agenda for Statistics
on the Social Economy

Marie J. BOUCHARD

Full professor, Université du Québec à Montréal, Canada

Damien ROUSSELIÈRE

Full professor, AGROCAMPUS OUEST, France

It is the authors who engage in a reflective questioning on the data and their construction who are more likely to receive criticism, doubt and suspicion from readers. By contrast, those who favor a non-reflective use of the data, maintaining the reader permanently at a very realistic level, protect themselves in advance from all criticism or ordinary doubt. [Our translation] (Lahire, 2006: 133)

This book brings together texts by authors who have directly or indirectly participated in the production of figures on the social economy across regions, countries and even internationally, usually with or for national statistical agencies and social economy actors. Compiled in chapters, their contributions offer an international perspective on these issues, presenting analyses from Belgium, Brazil, Canada (Quebec), Spain, the United States, France, Japan and the United Kingdom. The authors were asked not so much to elaborate on the methodologies used, which are already well described in the studies underlying the respective chapters, but to point out the contributions and limitations of the various approaches and to discuss their implications for public policy. The aim of this book was to open the black box of statistics in order to shed light on the scientific choices, understand the conditions of reproducibility and draw practical consequences. As emphasized by Pascal Rivière (2002: 145), this problem is largely underestimated by non-specialist users of this type of data: "Non-specialists might have the feeling that building statistics on businesses is a very simple task: it seems one just has to 'add facts'." We therefore thank the participating authors for agreeing to engage in this game of transparency, which is not without risk as pointed out by Bernard Lahire in the heading of this chapter. The compilation of chapters intends to help meet a real need, namely that of having an overview of the workings of statistical portraits on the social economy in order to inform researchers, stakeholders and policy-makers.

Overall, the book shows that the production of statistics on the social economy (and its variants) has become increasingly complex. It spans several dimensions, among them the methodological, theoretical and political dimensions, and also raises the questions we address in this conclusion: (1) Why is it important to have statistics on the social economy? and (2) How are social economy statistics produced? Throughout this conclusion, we point to the contributions and limitations of current methodologies and indicators and evoke some of the questions that remain to be clarified. We conclude (3) by summarizing some of the key recommendations that emerge from reading this book. These recommendations are intended for both researchers and public policy.

1. Why is it important to have statistics on the social economy?

Recent years have seen increased calls for a better understanding of the scope and impact of the social economy. These calls come from national and supra-national public bodies, which are showing a renewed interest in this form of economy. At the same time, they come from the actors of the social economy, who seek better visibility and recognition of their impacts. Although initial steps have been undertaken to meet this challenge, there still was no body of work that focused on evaluating the existing methods for identifying the size and weight of the social economy.

Today's public recognition of the social economy benefits from the capacity to show for its importance through various methodologies that were first developed by academics and further recognized and adopted by national and international instances, as it is the case for satellite account manuals for the non-profit institutions (see chapter by Salamon *et al.*) and for the cooperative and mutual organizations (see chapter by Fecher and Ben Sedrine-Lejeune), but also with the observatories and national surveys that have been coconstructed by social economy actors with the help of researchers and national statistical agencies (see chapters by Demoustier *et al.* and by Gaiger). Such works show that the social economy is present in nearly all industries and socio-economic contexts, and that it has an important contribution to the production of credence goods and collective services. As such, the social economy must, and can, play an important role in the search for innovative solutions to the challenges of sustainable development, more equitable distribution of wealth and territorial development, and other issues. That said, since interest in the measurement of the social economy is relatively recent, dating back only two or three decades, the codification of the field is yet to be perfected, especially in national statistics accounts. In fact, the practices and roles as well as the theoretical field of the social economy are multifaceted and ever-evolving (Hiez and Lavillunière, 2014; Draperi, 2012; Leroux, 2013). However, as daunting as these challenges may appear, they must nevertheless be taken on (DiMaggio, 2001) and do not, essentially, differ from those identified for over a decade with regard to new social processes or phenomena such as new information technologies or issues relating to sustainable development (Custance and Hillier, 1998; Jeskanen-Sunström, 2003).

While the social economy sector does account for a non-negligible portion of the overall economy – around 10% according to estimates[1] – it is its contribution to the overall regulation of the economy that makes it a significant player. For example, it serves as a mediator between producers and markets (e.g., agriculture); it extends the offer to less profitable demand segments (e.g., banking); it serves to counterbalance market power (e.g., food consumption, housing); it reduces information asymmetries (e.g., personal services); it democratizes the control and direction of businesses and territories (e.g., workers cooperatives, local development associations); and plays an innovative role in new activities (e.g., recovery and recycling, renewable energy). The social economy also ensures greater stability to the economic system by contributing to its smooth operation and by being able to counterbalance its fluctuations (Ansart, Artis and Monvoisin, 2014). It is also seen as a response to the crisis (Demoustier and Colletis, 2010; Draperi, 2011) or as an alternative to capitalism (Jeantet, 2008), or at least as a bulwark against the neoliberal ideology that has permeated not only the economy but society as a whole (Leroux, 2013).

However, it is difficult to advance this type of argument without seeking to provide empirical evidence, since the researcher on the social economy is, in the words of Pierre Bourdieu (2002), "neither a prophet nor a guru." The scientific understanding of the social economy and the validation of social economy theories therefore require reliable, longitudinal and reproducible data in order to be able to contribute to this theoretical debate. Among the research initiatives already taken in this direction, we refer to research showing how the creation of production cooperatives are tied to the economic and political cycles (Pérotin, 2006; Ingram and Simons,

---

[1] This is in the best-case scenarios and with actual methodological conventions.

1997) and to studies on the ability of agricultural cooperatives to survive (Nunez-Nickel and Moyano-Fuentes, 2004; Rousselière and Joly, 2012); on the effect of government subsidies on market income for non-profit organizations (Smith, 2007; Bouchard and Rousselière, 2011); and on the pluralist performance and growth of social economy enterprises (Backus and Clifford, 2013). What all of these works have in common is that they were able to draw on data from public sources and specific surveys.

The quest for a better understanding of the nature, role and place of the social economy in socio-economic regulations has been on the scientific and political agenda of the social economy for many years (Demoustier, 2014). In our approach to statistical production, we seek to build on this body of research and to identify the contributions, limitations and challenges of statistical production on the social economy at present.

2. How are social economy statistics produced?

The primary function of statistical measures on the social economy is to measure the participation and relative weight of this type of economy within the overall economy, in other words, its share in the gross domestic product (GDP), in employment and in the different sectors of activity. A statistical portrait is intended to represent the scope of a phenomenon, to highlight its main components and relative importance, to document some of its branches or subsectors, to follow its evolution over time, and, if possible, to allow for comparison with other phenomena. Measuring the size and scope of a segment of the economy involves defining and taking stock of the types of entities that are to be considered as belonging to that segment, and to measure their contribution to the economy (e.g., see Rivière 2002 for the small business survey). The first action entails counting the entities, and the second action requires measuring their added value. Although seemingly simple, these two actions pose particular challenges for the social economy.

2.1. Qualifying and classifying

The first task in any production of statistics is to define the "object" or the "entities" to be measured (Desrosières, 1993), namely by defining the rules for building the statistical population (see Baffour *et al.*, 2013, for a survey on the quality procedures for official statistics). Identifying the entities that make up the social economy generally involves three main tasks: 1) the identification of entities in the economic sectors and activity sectors most likely to contain social economy organizations; 2) the selection of entities by their legal status; 3) the selection, based on a set of qualification criteria, of the entities that qualify as belonging to the social economy according to the institutional (national) definition used, regardless of whether they have the legal status of being a social economy organization. The quality of this selection process depends on five different factors: the coverage of the studied population; the integrity (vitality) of the qualified entities; data availability; comparability of the study; and sustainability (continuity) of the study, which refers to its potential for replication. A classification system is considered robust when it is exhaustive enough to cover all entities; when it enables comparison; when it organizes the data along a unique and homogeneous dimension; when it provides mutually exclusive categories; and when it offers detailed granularity to enable refined analysis but also a structure to aggregate the data (Bouchard *et al.*, 2008).

Statistical studies about the social economy generally emphasize the primacy of the social purpose over the economic activity (chapter by Bouchard *et al.*). This primacy is evident from the empirical features that are typical of the organizational structures and operating modes of the social economy and that distinguish it from the rest of the economy. The resulting conceptual frameworks usually first establish which type of entities, legal statuses and activity sectors are excluded, and then identify a cluster of qualification criteria and statistical indicators for social economy organizations. Portraits capture either the whole of the social economy (e.g., when using national definitions (chapter by Demoustier *et al.*)), components (e.g., the nonprofit and voluntary sector or the cooperative and mutual society sector) or variations (e.g., the solidarity economy in Brazil (chapter by Gaiger)).

Another important component of a statistical portrait consists of classifying the entities into the proper group to allow correlating their respective contributions to specific types of activities. As explained in the chapter by Archambault, social economy entities, being either market or non-market producers, must be included in the international standard classifications systems, and they are – only not completely. Hence, there are advantages and drawbacks of these classifications. However, national accounting also has two precious and irreplaceable qualities. The first is that it enables comparison between different human activities and allows gauging them across common quantitative scales. The second is that, as a kind of universally accepted grammar, it allows to formalize, estimate and elucidate complex realities or poorly understood interdependencies.

2.2. *Limits and challenges of qualification and classification*

In the case of the social economy, one of the difficulties is that, apart from a few exceptions such as in France (chapter by Demoustier *et al.*), the national statistics systems offer no indicators with which to clearly identify or distinguish this subset of the economy. This is due in part to the fact that in most countries the social economy is still poorly codified within public policy (see chapter by Kurimoto). In addition, given the different origins, cultural and political customs and national institutional frameworks across countries, the social economy and its components operate with a different set of legal definitions, functions and behaviors in each country (DiMaggio and Anheier, 1990; Kyriakopoulos, 2000). This necessarily raises the issue of comparability. A further difficulty lies in the fact that the naming and qualification of social economy entities is not always consistent. For example, entities may be named depending on how the sector is named in the region in question (e.g., social economy in Europe, solidarity economy or popular economy in Latin America), their organizational structure and mode of operation (e.g., non-profit organization in the United States vs. association in France), or the values that drive them (certified vs. uncertified cooperatives in Belgium). In some cases, such as the solidarity economy in Brazil, which is characterized by a large number of informal enterprises, the production of portraits requires labor-intensive methodologies such as snowballing (chapter by Gaiger). A third difficulty is the permeable nature of the boundaries of the social economy, which is often seen to consist of "hard core" components yet also of "peripheral" (Desroche, 1983) or "hybrid" (chapter by Spear) components, the core having "porous borders" that can eventually be crossed by an organization during its lifespan.

Finally, emerging concepts such as those of the social enterprise, social business and social entrepreneur as well as the need for international comparability have prompted a search for the boundaries of the social economy beyond the legal statutory schemes traditionally associated with it (cooperatives, non-profit organizations or associations, mutual organizations and foundations). However, these concepts are still at a pre-paradigmatic stage (Nicchols, 2011). Moreover, to date there is no empirical evidence showing that there exists a distinction between non-statutory social enterprises and mainstream economy agents. In a recent study, researchers admitted that measuring and comparing social enterprise activity across Europe remains a challenge due to "the diversity of national economic structures, welfare and cultural traditions and legal frameworks" and to "the lack of availability and consistency of statistical information on social enterprises across Europe" (Wilkinson, 2014: 1), let alone across continents. Any solution to this challenge can only be found by means of robust methodologies.

2.3. Measuring the production

The chapters of this book present various methodologies that are applied to different segments or components of the social economy. Although the social economy takes various organizational forms and sets itself in different economic and institutional national settings, all the studies discussed in this work recognize it as a specific field to investigate in terms of national statistics. The main reason for this is that the social economy's goal is distinctive enough to make it worthwhile to sort it out from the rest of the economy. However, for the same reason, standard measuring methods tend to show those aspects of the social economy that are similar to those of the rest of the economy, for which the measurement tools had been first developed. While this enables comparison, it leaves out part of what is of particular importance to the social economy, namely the "social".

Statistics on the social economy show that it is a fully-fledged economic agent of the same order as other forms of the economy, such as the private sector or the institutional sector of non-financial businesses (see chapter by Artis *et al.*). In terms of outputs, social economy statistics reveal the contribution of this type of economy to the creation of wealth and identify its part in the various sectors of activity. As for inputs, the social economy was shown to provide a substantial portion of jobs for the active population in the different economies. This has prompted national accounting systems to enable the measurement of the non-market production of NPOs (see chapter by Salamon *et al.*). Such approaches also allow revealing the geographic distribution of the social economy through the quantification and localization of social economy activities taking place on the territories (in terms of number of businesses, volume of jobs and sectors of activity).

Apart from these structural and geographic representations, a further view of the social economy consists of looking at the resources it mobilizes. This involves, for example, showing the diversity and the interlocking of the various available financial resources, namely private market resources (sales), public quasi-market resources (contracts with government), non-market private resources (gifts) and public resources (subsidies). Observing such figures about the social economy then allows to confirm or rule out (as in the case of Brazil or Switzerland) the idea that the sector is essentially dependent on subsidies and to reaffirm the place of the social economy in the economic sphere. In many cases, however (as in Canada or France), the national statistical agency does not describe and itemize the different resources coming from a public source, listing them rather as subsidies. Yet, changes in public management, such as a reduction of subsidies and an increase of the calls for tenders or public services contracts, tend to alter the mode of interaction between the social economy and the public actor. In this matter, the exercise of producing social economy statistics functions to improve how the national agencies and the social economy actors themselves conceive the types of financial resources they raise to conduct their activities.

2.4. Limits and challenges in the measures of the production

Identifying the scope of activities with the usual means allows to see that the social economy represents, in a society like France for example, about 10% of the economy (National Observatory of ESS, 2014), with the public sector accounting for 20% and the capitalist economy for 70% (Leroux, 2013). Nevertheless, this 10% figure is somewhat inaccurate since volunteerism, even when measured in full-time equivalent positions (at minimum wage, replacement cost[2] or opportunity cost), could not be quantified in unpaid work since one aspect of what volunteers contribute is by nature "priceless." Apart from the statistical issues specific to its measurement (see Archambault and Prouteau, 2010), viewing volunteer work as nothing other than a resource undermines its role as creator of social ties within territories and as an indicator of citizen commitment undertaken by individuals (Demoustier, 2002). This raises the question of the measurement or the evaluation of the real contribution of the social economy.

Social economy statistics can hardly express the full range of value added generated by this economy. For example, collective forms of entrepreneurship and the internal governance of social economy organizations lead to the democratization of the economy in terms of accessibility, revenue distribution and economic countervailing power, among other aspects. Yet, such contributions are in general poorly taken into consideration. As well, progress is still to be made in understanding the spill-over effects and structuring impacts of social economy organizations in the different sectors. Other impacts also need to be assessed, calling for new types of indicators (chapter by Archambault). The chapter by Mertens and Marée exposes the difficulties to generate management indicators (profitability ratios, structure ratios, etc.) that correspond to enterprises not primarily aimed at generating profits, or to create statistics on a macroeconomic level referring to conventional measurements based on market indicators. Those often prove to be poorly adapted for providing an accurate quantitative understanding of what a social economy enterprise produces and of the impacts of social economy activities on people other than direct consumers. This topic is addressed for the case of cooperatives in the chapter signed by Uzea and Duguid.

What we find difficult to show about what the social economy produces is just as hard to show for the rest of the economy. Externalities,

---

[2] Based on the replacement cost method using observed market wages, the economic value of volunteer work throughout the world would, "if it were its own country, have the second largest adult population of any country in the world, and [...] be the world's seventh largest economy" (Salamon *et al.*, 2011: 217).

spillover effects, distributional effects, in short the "social" that is included in the economy is not well measured by classical statistical measurement of the economy and therefore has been the subject of an ongoing debate among statisticians and economists.[3] But the "social" is of crucial importance for the social economy as it is its *raison d'être*. Hence, whatever findings come from research on the subject should in the end serve both the social and the classical economy, shedding light on a larger spectrum of concerns about how the economy in general contributes – or not – to social wellbeing.

2.5. Satellite accounts and surveys

At present, statistics on the social economy are mainly produced with two approaches: satellite accounts, and observatories or survey studies. The satellite account is a methodology that is based on national statistical accounts. It has the advantage of using data that has already been entered and standardized and that is comparable to other fields of the economy. The methodology was initially developed for the social economy in Spain in 1995 (Barea and Monzón, 1995) and for the non-profit economy in the United States (Salamon *et al.*, 1995). Today, it is used in many countries (see chapter by Fecher and Ben Sedrine-Lejeune) and also allows to target specific segments of the social economy (see chapter by Monzón). One of the main advantages is of course the capacity to compare the weight of the social economy within the national accounts of a given country, but also from one country to another or from one period to the next. International comparison is facilitated in that the non-profit sector uses the statistical accounts manuals adopted from the United Nations in 2003, and in that the cooperative and mutual societies align themselves with the reports and manuals commissioned by the European Commission in 2006 (Barea and Monzón, 2006). The chapter by Monzón shows how the work of researchers led to the conceptual and statistical identity of the social economy, particularly in Europe. The chapter by Salamon *et al.* describes the various steps within a research program working toward the international comparability of the non-profit sector. Chapters by Demoustier *et al.* and by Gaiger also show how methodology is also a social construction and, in the cases discussed, developed through the deliberate consultation and active collaboration of researchers, statistical agencies and social economy actors. As shown with the comparative non-profit sector project, the project of measuring volunteer

---

[3]   See for example the *Handbook on Recent Developments in Ecological Economics* edited by Martinez-Alier and Ropke (2008) which provides a state-of-the-art review of this issue.

work (Salamon *et al.*, 1996; 2011) and the project of measuring a larger spectrum of the social economy (Barea and Monzón, 1995; 2006), such constructions can indeed contribute to the evolution of official statistical accounting methods and in particular to classifications (see chapter by Monzón).

2.6. Limits and challenges of satellite accounts and surveys

However, the satellite accounts method also inherits the limitations of national accounts (chapter by Artis *et al.*) with respect to providing classification nomenclatures (chapter by Archambault), the inventorying of small or hybrid entities (chapters by Bouchard *et al.* and by Spear) and the measurement of the real output of the social economy (chapter by Mertens and Marée). Observatories and surveys tend to gather specific data using non-standardized indicators which, although well-suited for the purposes of the social economy, are generally difficult to aggregate from one study to another unless a convention has been established among observers (chapter by Demoustier *et al.*). The dilemma is the indexicality, as defined by Garfinkel (1972), that is inherent in the social science enquiry. There is, first, the need for a specificity of the methods, tools and definitions in order to take into account the specificity of the social economy, at the risk of a lack of generalizability (between countries or over time). Conversely, there is also the need to develop generic tools (national accounts, surveys, etc.) in order to take account of the particularities of the social economy in each country. For example, a survey of cooperatives in a national business survey should systematically include questions about the membership to ensure the proper measurement of the cooperative sector. This dilemma is not unknown in public statistics (Desrosières, 2014) and is taken into consideration in compulsory surveys in agriculture by the implementation of recommendations of Eurostat.[4] Still, the dilemma persists, namely because the procedures existing in public bodies foster inertia and allow only for minimal development. Yet, the replicability of surveys and the comparability of results for the social economy could be advanced by means of quality procedures. Such procedures exist for

---

[4]   Eurostat aimed to align the practices of national institutions as much as possible with the successive waves of agricultural censuses so that generic results could be compared across countries while also allowing certain elements to the discretion of the countries, given the unique characteristics of their respective national economies. France, for example, opted to pose additional questions to wine cooperatives and farm machinerie cooperatives; and these questions were absent in the surveys conducted in the other countries. Similarly, emerging phenomena (such as short food supply chains) are addressed in different ways depending on the country (e.g., percentage of sales in France, agricultural diversification practices in Ireland).

the satellite accounts and are made available through the publication of manuals. However, in the case of surveys or observatories on the social economy, such manuals are, for the most part, yet to be created. Recent research on quality procedures with regard to official statistics (e.g., Blasius and Thiessen, 2012) could be inspiring in this regard.

3. How might we better understand the social economy in the future?

The work presented in this book confirms the need for statistics on the social economy, yet also that researchers and national statistical agencies are well on their way of meeting that need in an efficient manner. Through various methodologies, numbers and figures are being produced that depict and weigh the contribution of the social economy. Nevertheless, future research is required to address issues relating to three general findings emerging from this book. The first is that the methods for measuring the social economy (and its variants) differ greatly from one study to another, but that each sheds light on specific dimensions of the social economy. The second finding is that, despite the social economy being an important economic actor, a part of its contribution and output remains in the shadow of statistical measurement, which raises questions about the ability of statistical tools to properly grasp the social economy. The third finding is that quantification – and the different ways to quantify – has a transforming effect on the definition of the social economy and contributes to institutionalize it. These dimensions are interconnected and the status of one affects the other two. There is thus a rebound effect between the definition, measurement and public recognition of the social economy.

The chapters in this book highlight well the many questions raised by the issue of the heterogeneity of the social economy, which concerns above all qualification and classification. In general, the categories used should be wide enough to be relevant and generalizable at the aggregate level but at the same time be sufficiently restrictive to reflect the reality at the level of organizations and their activities (chapters by Archambault; Bouchard *et al.*; Salamon *et al.*; Mertens and Marée). Heterogeneity moreover poses purely empirical challenges concerning identification, namely that of recognizing the social economy beyond its diverse legal statutes (chapters by Gaiger and by Spear) and administrative databases (chapters by Artis *et al.* and by Kurimoto); and of identifying, within a given legal status, which entities take part in the social economy (chapter by Fecher and Ben Sedrine-Lejeune). These points have been addressed in other contributions. For example, the unanticipated *ex ante* use by certain social groups of very flexible legal forms, such as medical cooperatives in South Korea, poses the problem of an *ex post* identification of those coming from the social economy (see Bidet and Eum, 2014). We know that this type of ad hoc mode of identification (e.g., cooperatives that are on the list of a federation) is problematic and requires significant work on the quality of the data. This may be due to the fact that the owners of lists, whose influence may be determined, as in federations, by the number of members, have no direct interest in purging the lists of possible errors. Our own work on the Montreal social economy (Bouchard *et al.*, 2008), which is based on this type of approach, revealed the existence of different types of errors, such as the presence of "ghosts" or long-missing organizations, organizations listed multiple times and organizations listed by mere declaration. It is this lack of control over this problem by researchers, and less the sampling method, that can lead to significant biases.

We thus observe strengths and weaknesses of the available tools in aligning the portraits of the social economy to the national statistics and, vice versa, limits of national statistics to inform effectively about the social economy. For example, the chapter on Brazil (by Gaiger) raises the challenge which the informal economy represents as a part of the social economy (solidarity), while the chapter on the United Kingdom (by Spear) points to the various forms of hybridization that ultimately lead to the disappearance of borders between key sectors of the private, public and social economy. While in the past, the issue was to find ways to quantify non-market activities, today's problems have to do with the weakening frontiers between legal statutes. Other chapters focus on the need to develop or improve indicators to allow for a better detection of the presence of the social economy in national statistics (chapter by Archambault), a better assessment of the ability of the social economy to respond to unmet needs and stabilize local economies (chapter by Uzea and Duguid), and a better measurement of the non-market dimensions and the impacts of social economy activities (chapter by Mertens and Marée).

The book also discusses population dynamics and the challenge of obtaining reliable data to monitor the social economy over time, knowing that the frontiers of the social economy, and its definitions, can evolve in response to political events and trends (chapter by Spear). Researchers can also be expected to continually adjust their methods to take into account the new dimensions of the social economy. Given that such adjustments may threaten the longitudinal consistency of methods (chapter by Salamon *et al.*), the latter should be designed so as to be reusable (chapters by Gaiger and by Kurimoto), which requires transparency in their administration.

As mentioned in the chapter by Artis *et al.*, one challenge is to quantify the economic weight of this economy in a way that is internationally comparable. Another challenge, in seeming contradiction to the latter, is to successfully convey the aspects of this type of economy that are not economic in the strict sense of the term, and the role it plays in the different contexts where it takes root. This is important in order to prevent a simplistic reading of the social economy as nothing other than a government instrument or a commonplace market player (DiMaggio and Powell, 1983). The methodological questions that then emerge concerning social economy statistics have to do with: 1) the quality of standard indicators compared to specific indicators (chapters by Archambault, and by Mertens and Marée); 2) the advantages of detailed approaches that are embedded in specific and local realities versus broader approaches that allow for comparison with the rest of the economy and with national social economies of other regions (chapter by Demoustier *et al.*); and 3) cost-benefit ratios of different methodologies on the basis of targeted objectives (chapter by Bouchard *et al.*).

Finally, the issues concerning quantification are not unrelated to those of the institutionalization of the social economy. The case of Japan (chapter by Kurimoto) is instructive because it shows that a state control that is too strong has negative consequences for the recognition of the social economy, including in national statistics. In the UK (chapter by Spear), applying the definitional criteria differently has lead to question the extent of growth in the sector – as changes to the operational criteria indeed pose considerable challenges to longitudinal analysis of performance (Teasdale *et al.*, 2013). The chapter on France (by Demoustier *et al.*) highlights how the challenge of producing social economy statistics may differ depending on the developmental phase of a society, the alliances in place, the political parties in power and the diversity of levels (European, national and local). At present, the definition of the social economy advanced in most studies combines a selection of legal statuses (cooperatives, non-profit and mutual organizations, foundations) and of sectors of activities. Some actors, however, suggest to exclude large commercial cooperatives and mutual organizations, restricting the scope to non-profit and limited-profit-distribution entities.[5] Others promote the use of a much wider spectrum that tend to put statutory social economy enterprises on par with "socially responsible" privately held for-profit businesses. Some even forego a more detailed specification altogether and, making do with more vague notions of "social enterprise" or "social entrepreneurship," express a preference for the blurriness of "a fluid institutional space" (Nicchols, 2011: 612). This shows how the topic of social economy statistics is, still

---

[5]   See the definition proposed by the Third Sector Impact research program: http://third-sectorimpact.eu/documentation/first-tsi-policy-brief-defining-third-sector/.

today, part and parcel of the debate regarding the institutionalization of this kind of economy. Part of that debate, as Nicchols (2011) points out, involves social actors that are external to the field of the social economy – such as large foundations and business faculties – hence marginalizing those who should be first concerned by defining their own action and structures.

4. Concluding remarks

It is important to note that this book does not aim to be exhaustive with regard to social economy statistics but rather to provide an overview of our knowledge and capabilities to produce them, and of some of the methodological challenges that lie ahead. Note that in some countries the concept of the social economy does not exist or that its components do not recognize each other as being part of a greater whole. For example, in the United States, the concept of social enterprise refers to the voluntary and non-profit sector and to its new entrepreneurial turn,[6] the concept of the solidarity economy refers to citizen initiatives that resist capitalism,[7] and cooperatives and mutuals are grouped into two different national associations.[8] Or, in certain other countries, such as in Korea or parts of Eastern Europe, the presence of large government-led cooperatives is considered by some researchers to account for a "pseudo social economy" (Bidet, 2009; Monzón and Chaves, 2012). Finally, there are countries such as Australia, where the notion of the social economy was initially associated solely with the non-profit sector and only later expanded to cover a wide range of businesses that subscribe primarily to social objectives and whose surpluses are principally reinvested for those objectives.[9]

Another aspect that was not addressed in this book concerns secondary studies (i.e., studies that examine what is actually done with the available data). As becomes evident from a review of scientific journals, this field of applied statistics is currently very dynamic. Econometric methods, the application of which has generally been limited to the analysis of linear and one-dimensional phenomena (Van Staveren, 1999), are now more likely than ever to consider the specificities of the social economy. These methods enabled, for example, the analysis of the heterogeneity of the social economy through latent class analysis (Rousselière and Bouchard,

---

[6] See definition of the US Social Economy Network: http://www.socialeconomynetwork.org/.

[7] See the US Solidarity Economy Network: http://ussen.org/.

[8] See National Cooperative Business Association: http://www.ncba.coop/; and National Association of Mutual Insurance Companies: http://www.namic.org/.

[9] See Finding Australia's Social Enterprise Sector: http://www.socialeconomy.net.au/.

2011; Hustinx *et al.*, 2014) and quantile regression methods (Christensen, 2004; Clemente *et al.*, 2012); the consideration of the plurality of objectives targeted in the multi-output models (Becchetti and Pisani, 2015); and the consideration of cross-causal effects through simultaneous equations (Pennerstorfer and Weiss, 2012). Moreover, the growing interest in mixed research approaches or cross-over research in general (Small, 2012), combining qualitative and quantitative analyses, also emerges as a very promising path to follow. For example, the work of Francesconi and Ruben (2014) combining qualitative analysis (focus group) and quantitative longitudinal analysis allows to well justify the statistical choices and interpret the results of an impact analysis of fair trade on the performance of cooperatives.

Certain avenues of research in this area that are promising but not yet applied to the social economy may also face the limitations of some current practices pointed out in our book. In that context, we recommend using tools such as recent Bayesian methods of contingent valuation, which introduce reflexivity in the analysis and take into account the measurement errors (Balcombe *et al.*, 2007), as well as multi-agent or micro-simulation methods going beyond former input-output models (Diaz, 2011; Lennox and Armsworth, 2013) when assessing the value and impact of the social economy. In terms of the survey methodology, we recommend mixed and participatory methods developed in public statistics (Willis *et al.*, 2014).

This book is to be regarded as a starting point for future studies and to provide an overview of the currently large variety of approaches and questionings posed by this unwieldy topic of research. As such, the book represents an invitation to address some of these challenges in order to overcome the limitations of known approaches. Ultimately, the book offers two types of recommendations: one for public authorities and one for researchers. The production of social economy statistics is as much a scientific question as a political one in the sense that statistical results are decision-making tools for public and private actors. Yet, quantification techniques have been subject to criticism since their inception (Desrosières, 2014: 62). One of the reasons is that

> quantification transforms and stiffens the measured realities. It supposes that a series of equivalence conventions be developed and clarified ahead, involving comparisons, negotiations, compromises, translations, inscriptions, codifications and codified and replicable procedures and calculations leading to numbers. Measurement as such only comes later, as the settled implementation of these conventions [our translation] (Desrosières, 2014: 38).

The social economy is now officially recognized in many legislatures, which should prompt statistical agencies to develop new or adapt existing indicators to capture its reality. As already pointed out, a survey of the social economy incurs significant costs, since it involves taking into account more aspects than is required for other types of surveys. Thus, ensuring that all aspects are included in official statistics comes at a price. The development of statistics by public institutions must be advanced and will most likely require the initial support of researchers before becoming decentralized (among actors and specialized researchers). This includes finding room in the mandatory questionnaires and surveys of national statistical agencies to make the relevant data of interest to the sector visible and accessible.

Another topic that merits further attention is that of professional ethics for researchers, who, as data producers, are stakeholders of the field they are studying. This is an important dimension in the social economy. Any production of statistical data that claims objectivity must follow a code of ethics (see the *International Statistical Institute's Declaration of Professional Ethics* discussed in Biemer and Lyberg, 2003: 375). Among other recommendations, any search for avenues to pursue in social economy research should take account of the importance of conducting sensitivity tests (now often absent), as is increasingly done in other areas (e.g., medicine, political science) (Ince, 2012; Gelman, 2013) in order to see if results are robust to alternative methodological choices.[10] There is also a need to open the black box concerning the methods used, although this may seem risky, as underlined in many works by Desrosières and Lahire. Our book shows a strong gain in transparency in the implementation of surveys. Finally, it also shows the challenge of rendering research more reproducible, namely through the provision of data and methodologies and through a clear presentation of the retained hypotheses.

Statistics are not the only way to take measure of the social economy and promote its recognition. Other methods exist across organizations and sectors to assess the output of the social economy in correspondence with public policies, business strategies or values within the social economy movement (see Bouchard, 2009). Networks, clusters, federations and the larger ecosystem around the social economy are also important in enhancing the visibility and legibility of the sector. Nevertheless, statistical representation remains a necessity if we wish to understand the social economy as a whole and to ascertain its boundaries, organizational forms and contribution to the economy with a view to establishing measures that are comparable from one country or region to another as well as across

---

[10] For example, the impact of non-responses on the final results can be examined according to various assumptions concerning the scope or using multiple imputation techniques (Van Buuren, 2012). The existence of measurement errors due to social desirability phenomena in the replies (strategic behavior of respondents) can also be tested (Blackwell *et al.*, 2015).

time. This becomes all the more important in that governments increasingly call on economic actors to participate in the production of goods and services that have a social utility.

This book provides valuable lessons for that matter. Chapter by Bouchard *et al.* document an efficient approach to the identification of a statistical population from administrative databases, screening methods that will be of interest for every new endeavor to produce official statistics on social economy. Both chapters form Archambault, and Mertens and Marée highlight the conventional foundation of measurement aggregate economic estimates, an issue that feeds an ongoing debate in macroeconomics. Following Heckman (2005), Uzea and Duiguid show that no causal inference investigation can be a-theoretical. Therefore researchers have to make clear their theoretical assumptions. In effect, contrary to the underlying assumption of data mining, or "data fishing" (Lovell, 1999), that pretend to "let data speak for itself," chapter by Salamon *et al.* warns us about the fact that *"Numbers do not speak for themselves"*. Furthermore the chapters from specific studies on Belgium (Fecher and Ben Sedrine-Lejeune), Brazil (Gaiger), United Kingdom (Spear), Japan (Kurimoto) and France (Demoustier *et al.*) highlight the fact that public statistics made on a societal issue must involve a large number of partners: public authorities, social movement, academics… and not only statistical institute.

We believe that efforts should be made to improve the statistical representation of the social economy in national statistics and to develop observatories and conduct surveys. This should be done in concert with researchers and social economy actors to ensure the scientific rigor and legitimacy of this production among all stakeholders. Findings from research on the subject will not only serve the social economy but it should also be of use to studies that intend to measure social impacts of economic activities on a broader spectrum. This would moreover help demystifying the social economy to policy makers as well as to external stakeholders.

References

Ansart, S., Artis, A. and Monvoisin, V., "Les coopératives: agent de régulation au coeur du système capitaliste?," *La Revue des Sciences de Gestion*, 2014/5, Vol. 269-270, pp. 111-119.

Archambault, E. and Prouteau, L., "Un travail qui ne compte pas? La valorisation monétaire du bénévolat associatif," *Travail et Emploi*, Vol. 24, 2010, pp. 57-67.

Backus, P. and Clifford, D., "Are Big Charities Becoming More Dominant? Cross-sectional and Longitudinal Perspectives," *Journal of the Royal Statistical Society*, Series A (Statistics in Society), Vol. 176, No. 3, 2013, pp. 761-776.

Baffour, B., King, T. and Valente, P., "The Modern Census: Evolution, Examples and Evaluation," *International Statistical Review*, Vol. 81, No. 3, 2013, pp. 407-425.

Barea, J., and Monzón, J. L., *La Cuenta Satélite de la Economía Social en España: Una Primera Aproximación*, Valencia, CIRIEC-España, 1995.

Barea, J. and Monzón, J. L. (eds.), *Manual for Drawing up the Satellite Accounts of Companies in the Social Economy: Co-operatives and Mutual Societies*, Brussels, European Commission, D.G. for Enterprise and Industry and CIRIEC, 2006.

Becchetti, L. and Pisani, F., "The Determinants of Outreach Performance of Social Business: An Inquiry on Italian Social Cooperatives," *Annals of Public and Cooperative Economics*, Vol. 86, No. 1, 2015, pp. 105-136.

Bidet, É., "La difficile émergence de l'économie sociale en Corée du Sud," *RECMA, Revue internationale de l'économie sociale*, Vol. 310, 2009, pp. 65-78.

Bidet, É. and Eum, H., "Nouvelles formes de protection sociale: Entreprises sociales et coopératives médicales en Corée du Sud," *Revue française des affaires sociales*, Vol. 3, 2014, pp. 84-97f.

Biemer, P. P. and Lyberg, L. E., *Introduction to Survey Quality*, Hoboken, Wiley, Wiley Series in Survey Methodology, 2003.

Blackwell, M., Honaker, J. and King, G., "A Unified Approach to Measurement Error and Missing Data: Overview," *Sociological Methods & Research*, 2015, forthcoming.

Blasius, J. and Thiessen, V., *Assessing the Quality of Survey Data*, London, Sage Publications, 2012.

Bouchard, M. J., *The Worth of the Social Economy, An International Perspective*, Brussels, Peter Lang, CIRIEC collection Social Economy and Public Economy, 2009.

Bouchard, M. J., Ferraton, C., Michaud, V. and Rousselière, D., *Base de données sur les organisations d'économie sociale. La classification des activités*, Montreal, Université du Québec à Montréal, Chaire de recherche du Canada en économie sociale, Collection Recherche no R-2008-1, 2008.

Bourdieu, P., *Interventions (1961-2001). Sciences sociales et action politique*, Marseille, Agone, 2002.

Christensen, E. W., "Scale and Scope Economies in Nursing Homes: A Quantile Regression Approach," *Health Economics*, Vol. 13, 2004, pp. 363-377.

Clemente J., Diaz-Foncea, M., Marcuello, C. and Sanso-Navarro, M., "The Wage Gap Between Cooperative and Capitalist Firms: Evidence from Spain," *Annals of Public and Cooperative Economics*, Vol. 83, No. 3, 2012, pp. 337-356.

CNCRES (Conseil National des Chambres Régionales de l'Economie Sociale) and Observatoire national de l'ESS, *Atlas commenté de l'économie sociale et solidaire*, 2014.

Custance, J. and Hillier, H., "Statistical Issues in Developing Indicators of Sustainable Development," *Journal of the Royal Statistical Society*, Series A (Statistics in Society), Vol. 161, No. 3, 1998, pp. 281-290.

Demoustier, D., "Le bénévolat, du militantisme au volontariat," *Revue française des affaires sociales*, 2002, Vol. 4, No. 4, pp. 97-116.

Demoustier, D., "Le rôle régulateur visible et invisible des coopératives," in *Le pouvoir d'innover des coopératives. Québec 2014. Sommet international des coopératives*, 2014. www.sommetinter.coop/files/live/sites/somint/files/pdf/ Articles%20scientifiques/2014_04_Demoustier.pdf.

Demoustier D. and Colletis, G., "L'économie sociale et solidaire face la crise: simple résistance ou participation au changement?," *RECMA, Revue internationale de l'économie sociale*, Vol. 325, 2010, pp. 21-35.

Desroche, H., *Pour un traité d'économie sociale*, Paris, Coopérative d'information et d'édition mutualiste, 1983.

Desrosières, A., *La politique des grand nombres: histoire de la raison statistique*, Paris, Éditions La Découverte & Syros, 1993 (ed. 2010).

Desrosières, A., *Prouver et gouverner. Une analyse politique des statistiques publiques*, Paris, La Découverte, 2014.

DiMaggio, P. J., "Measuring the Impact of the Nonprofit Sector on Society Is Probably Impossible but Probably Useful. A Sociological Perspective," in P. Flynn and V. A. Hodgkinson (eds.), *Measuring the Impact of the Nonprofit Sector*, New York, Kluwer Academic/Plenum Publishers, 2001, pp. 249-270.

DiMaggio, P. J. and Anheier, H. K., "The Sociology of Nonprofit Organizations and Sectors," *Annual Review of Sociology*, Vol. 16, 1990, pp. 137-159.

Draperi, J.-F., *L'économie sociale et solidaire, une réponse à la crise? Capitalisme, territoires et démocratie*, Paris, Dunod, 2012.

Francesconi, G. N. and Ruben, R., "Fair Trade's Theory of Change: An Evaluation Based on the Cooperative Life Cycle Framework and Mixed Methods," *Journal of Development Effectiveness*, Vol. 6, No. 3, 2014, pp. 268-283.

Gelman, A., "Ethics and Statistics: Is It Possible to Be an Ethicist Without Being Mean to People?," *Chance*, Vol. 26, No. 4, 2013, pp. 52-55.

Heckman, J., "The Scientific Model of Casuality," *Sociological Methodology*, Vol. 35, No. 7, 2005, pp. 1-97.

Hiez, D. and Lavillunière, É. (eds.), *Théorie générale de l'économie sociale et solidaire*, Luxembourg, Larcier, 2014.

Hustinx, L., Verschuere, B. and De Corte, J., "Organisational Hybridity in a Post-Corporatist Welfare Mix: The Case of the Third Sector in Belgium," *Journal of Social Policy*, Vol. 43, No. 2, 2014, pp. 391-411.

Ince, D., "The Problem of Reproductibility," *Chance*, Vol. 25, No. 3, 2012, pp. 4-7.

Jeantet, T., *L'économie sociale. Une alternative au capitalisme*, Paris, Economica, 2008.

Jeskanen-Sundström, H., "ICT Statistics at the New Millennium – Developing Official Statistics – Measuring the Diffusion of ICT and its Impact," *International Statistical Review*, Vol. 71, No. 1, 2003, pp. 5-15.

Kyriakopoulos, K., *The Market Orientation of Cooperative Organizations. Learning Strategies for Integrating Firm and Members*, Breukelen, The Netherland Institute for Cooperative Entrepreneurship, 2000.

Lahire, B., *La culture des individus*, Paris, La Découverte / Poche, 2006.

Lennox, G. D. and Armsworth P. R., "The Ability of Landowners and Their Cooperatives to Leverage Payments Greater Than Opportunity Costs from Conservation Contracts," *Conservation Biology*, Vol. 27, No. 3, 2013, pp. 625-634.

Leroux, A., *L'économie sociale. La stratégie de l'exemple*, Paris, Economica, 2013.

Lovell, M. C., "Data Mining," in S. Kotz, Read, C. B. and Banks D. L., *Encyclopedia of Statistical Sciences*, Thousand Oaks, Wiley, 1999, pp. 1567-1568.

Martinez-Alier, J. and Ropke, I. (eds.), *Recent Developments in Ecological Economics*, Cheltenham, Edward Elgar, 2008.

Monzón, J. L. and Chaves, R., *The Social Economy in the European Union*, Brussels, European Economic and Social Committee, 2008 and 2012.

Nicchols, A., "The Legitimacy of Social Entrepreneurship: Reflexive Isomorphism in a Pre-Paradigmatic Field," *Entrepreneurship Theory and Practice*, Vol. 34, No. 4, 2010, pp. 611-633.

Nunez-Nickel M. and Moyano-Fuentes J., "Ownership Structure of Cooperatives as an Environmental Buffer," *Journal of Management Studies*, Vol. 41, No. 7, 2004, pp. 1131-1152.

Ortega Diaz, A., "Microsimulations for Poverty and Inequality in Mexico Using Parameters from a CGE Model," *Social Science Computer Review*, Vol. 29, No. 1, 2011, pp. 37-51.

Pennerstorfer, D. and Weiss, C. R., "Product Quality in the Agri-Food Chain: Do Cooperatives Offer High-Quality Wine?," *European Review of Agricultural Economics*, Vol. 40, No. 1, 2012, pp. 143-162.

Rao, J. N. K., "On Measuring the Quality of Survey Estimates," *International Statistical Review*, Vol. 73, No. 2, 2005, pp. 241-244.

Rivière, P., "What Makes Business Statistics Special?," *International Statistical Review*, Vol. 70, No. 1, 2002, pp. 145-159.

Rousselière, D. and Bouchard, M. J., "À propos de l'hétérogénéité des formes organisationnelles de l'économie sociale: isomorphisme vs écologie des organisations en économie sociale," *Canadian Review of Sociology/Revue Canadienne de Sociologie*, Vol. 48, No. 4, 2011, pp. 414-453.

Rousselière, D., and Joly, I., "À propos de la capacité à survivre des coopératives agricoles: Une étude de la relation entre âge et mortalité des coopératives agricoles françaises," *Revue d'études en agriculture et environnement/Review of Agricultural and Environmental Studies*, Vol. 92, No. 3, 2012, pp. 259-289.

Salamon, L. M. and Anheier, H. K., "Social Origins of Civil Society: Explaining the Nonprofit Sector Cross-Nationally," Comparative Nonprofit Sector Working Paper #22, USA, The Johns Hopkins Center for Civil Society Studies, 1996.

Salamon, L. M. and Sokolowski, W., "The Third Sector in Europe: Towards a Consensus Conceptualization," TSI Working Paper Series No. 2, Brussels, Third Sector Impact, 2014.

Salamon, L. M., Sokolowski, S. W. and Haddock, M. A., "Measuring the Economic Value of Volunteer Work Globally: Concepts, Estimates, and a Roadmap to the Future," *Annals of Public and Cooperative Economics*, Vol. 82, No. 3, 2011, pp. 217-252.

Small, M. L., "How to conduct a mixed methods study: Recent trends in a rapidly growing literature," *Annual Review of Sociology*, 37, 2011, pp. 57-86.

Smith, T. M., "The Impact of Government Funding on Private Contributions To Nonprofit Performing Arts Organizations," *Annals of Public and Cooperative Economics*, Vol. 78, No. 1, 2007, pp. 137-160.

Starr, M. A., "Qualitative and mixed-method research in economics: Surprising growth, promising future," *Journal of Economic Surveys*, Vol. 28, No. 2, 2014, pp. 238-264.

Teasdale, S., Lyon, F. and Baldock, R., "Playing with Numbers: A Methodological Critique of the Social Enterprise Growth Myth," *Journal of Social Entrepreneurship*, Vol. 4, No. 2, 2013, pp. 113-131.

Van Buuren, S., *Flexible Imputation of Missing Data*, Boca Raton, Chapman & Hall/CRC, 2012.

Van Staveren, I., "Chaos Theory and Institutional Economics: Metaphor or Model?," *Journal of Economic Issues*, Vol. 33, No. 1, 1999, pp. 141-167.

Willis, G. B., Smith, T. W., Marco-Shariff S. and English N., "Overview of the Special Issue on Surveying the Hard-to-Reach," *Journal of Official Statistics*, Vol. 30, No. 2, 2014, pp. 171-176.

Wilkinson, C., *A Map of Social Enterprises and their Eco-Systems in Europe. Executive Summary*, Report submitted by ICF Consulting Services, Brussels, European Union, 2014.

Presentation of the Authors

Édith Archambault graduated from Université Paris 1 Panthéon-Sorbonne in economics, sociology and political science and has a doctorate in economics. Her academic career took place mainly at Université Paris 1 Panthéon-Sorbonne, where She chaired the Laboratoire d'Économie Sociale and was the Chair of the department of economics. Since 2004, she is emeritus professor. Her research fields are national accounting and the economics of the non-profit sector. She was the French local associate of the Johns Hopkins Non-profit Sector Comparative Project, and then expert for two handbooks on non-profit sector measurement (UNO and ILO). In 2010, she published the report *Knowledge of Nonprofits* for the Conseil national de l'information statistique (Cnis). Édith Archambault is the author of more than 250 publications and is an Officer of the Order of Légion d'Honneur and of the National Order of Merit.

Amélie Artis, PhD, is associate professor of economics at Sciences Po Grenoble. She is director of the master's program in social economy and researcher at PACTE (Politiques publiques, ACtion politique, Territoires), a research laboratory of the Centre national de la recherche scientifique (CNRS). Her research includes organization studies (cooperatives and associations), the history of economic thought, financial and banking questions and local development. She analyzes the evolution of cooperatives in the economic system, namely between continuity and renewal, using historical and institutional approaches. In 2013, she published *Introduction à la finance solidaire* (Presses Universitaires de Grenoble).

Wafa Ben Sedrine-Lejeune graduated from the Institut des Hautes Études Commerciales de Tunis in 1991. She obtained in 1993 a master's degree in economics at the Université de Liège, Belgium. She worked as senior researcher at the Université de Liège and as a teacher in a high school. Her research works and publications mainly concern health economics, public health, and cooperative economics.

Marie J. Bouchard is full professor at Université du Québec à Montréal, Canada. She holds a doctorate in sociology from the École des Hautes Études en Sciences Sociales, France. She is a regular member of the Centre de recherche sur les innovations sociales (CRISES), where she currently leads a research area on social innovations and collective enterprises. She was co-head of the Community Housing Research Partnership of the Community-University Research Alliance on the Social Economy (CURA-SE) (2000-2010). She held the Canada Research Chair on the Social Economy (2003-2013), where she dedicated her work to the conceptual and statistical marking of the social economy, social innovation and evaluation. In 2015, she was appointed president of the CIRIEC International Scientific Commission on the "Social and Cooperative Economy". In recent years, she published *Se loger autrement au Québec* (Éditions St-Martin, 2008), *The Worth of the Social Economy* (Peter Lang, 2009) and *L'économie sociale* *au Québec, vecteur d'innovation* (Presses de l'Université du Québec, 2011). An English version of the latter book was also published under the title *Innovations in the Social Economy, The Québec Experience* by University of Toronto Press (2013).

Élisa Braley was head of the Observatoire national de l'Économie Sociale et Solidaire at the Conseil national des chambres régionales de l'économie sociale et solidaire (CNCRESS) in France. She was the scientific editor of the *Atlas commenté de l'économie sociale et solidaire 2012*. Structured into six major themes (business demographics, actors, economy, territories, sectors of activity, and challenges and changes), the Atlas allows for a comprehensive understanding of what the social and solidarity economy is in France and offers an overview of its evolution.

Paulo R. A. Cruz Filho is a PhD student in administration at Université du Québec à Montréal, Canada. His research is dedicated to strategic management and the social economy. He is a professor of strategy at FAE Business School (Brazil), where he coordinates two postgraduate courses on social entrepreneurship and sustainability. He worked at the Canada Research Chair on the Social Economy from 2008 to 2013, where he co-authored a conceptual framework for defining the statistical population of the social economy in Quebec. He is an active research member of the International Comparative Social Enterprise Models (ICSEM) project, which aims at comparing social enterprise models and their respective institutionalization processes across the world. He is also a member of RILESS (Red de Investigadores Latinoamericanos de Economía Social y Solidaria), a Latin American network of researchers on the social and solidarity economy.

Danièle Demoustier is an economist at Sciences Po Grenoble, France, where she was director of the postgraduate studies module (DESS) "Politique de développement des activités et entreprises d'économie sociale." As part of her career, she contributed to numerous training seminars for social economy enterprises and organizations and is engaged in their general assemblies on a regular basis. She is a member of the Équipe de Socio-Économie Associative et Coopérative and of the European team of Création de richesses en contexte de précarité (CRCP) and has authored numerous publications on the social and solidarity economy. In 2003, she published her book *L'économie sociale et solidaire. S'associer pour entreprendre autrement* with La Découverte. In 2012, she collaborated with Nadine Richez-Battesti and Jean-François Draperi on the publication of the *Atlas national de l'économie sociale et solidaire*. Together with Rafael Chaves, she also co-edited the book *The Emergence of the Social Economy in Public Policy / L'émergence de l'économie sociale dans les politiques publiques*, published in 2013 by Peter Lang in the Social Economy & Public Economy series.

Fiona Duguid has a PhD from the University of Toronto in adult education and community development. She worked for the Rural and Co-operatives Secretariat of the Government of Canada as a senior policy and research analyst, followed by a position as research officer at the Canadian Co-operative Association. Her present research focuses on cooperatives, the social economy, sustainability and community economic development. She currently coordinates the National Study on the Impact of Co-operatives (Canada), the New Co-operative Development in Canada study, and a research project on women's cooperatives in Turkey funded by the World Bank. In addition she contributed to a recent cooperative sustainability study by the International Co-operative Alliance. She is on the board of the Canadian Association for the Study of Co-operation (CASC), the management committee of the Canadian Co-operative Research Network, and the Measuring the Co-operative Difference Network. An instructor with the Saint Mary's University Co-operative Business Management program, she currently also serves on the advisory committee for the Ottawa Renewable Energy Co-operative.

Fabienne Fecher became full professor in the Faculty of Social and Human Sciences of Université de Liège in 1996. She holds a master's degree in economics from Université Catholique de Louvain and obtained a PhD in economics, with a thesis entitled "Productivity, efficiency and technical change. Empirical studies," from Université de Liège in 1991. She teaches general economics, national accounting, social policy and health economics. She is scientific director of the Belgian section of CIRIEC, editor of the *Annals of Public and Cooperative Economics* (Wiley-Blackwell), member of the editorial board of the *Revista de Economia publica, social y cooperativa* (*CIRIEC-España*), member of the reading committee of the *Social Economy and Public Economy* series of Peter Lang Publishing Group, and member of the editorial board of the *Economie publique et économie sociale* series of Presses universitaires de Rouen et du Havre. Her fields of research are health economics (performance of health systems) and social economy (evaluation and public policies).

Luiz Inácio Gaiger has a PhD in sociology and is a researcher for the Brazilian National Research Council (CNPq) as well as a full professor of social sciences at the Universidade do Vale do Rio dos Sinos (Unisinos), São Leopoldo, Brazil. He has academic experience in citizen participation, social movements, solidarity economy and social entrepreneurship. He coordinates, with Prof. José Luis Coraggio (UNGS/Argentina), RILESS (Red Latinoamericana de Investigadores en Economía Social y Solidaria) and is co-director of the journal *Otra Economía*. Some of his recent publications are *Dicionário Internacional da Outra Economia* (edited with Antonio Cattani, Jean-Louis Laville and Pedro Hespanha, 2009, Coimbra: Almedina) and *A Economia Solidária no Brasil: uma análise de dados nacionais* (edited in collaboration with the *Ecosol* Research Group, 2014, São Leopoldo: Oikos).

Thomas Guérin is project coordinator at CRESS PACA (Chambre Régionale de l'Economie Sociale et Solidaire Provence-Alpes-Côte d'Azur). He holds a master's degree in social economy from Université de Marseille. His work focuses on the analysis of the specificities of social economy organizations. With the continuous support of Nadine Richez-Battesti and François Rousseau, he conceived and established the Observatoire régional and participated in the evolution of the national statistical methodology used by INSEE as well as in various action researches, in particular on new indicators of wealth. In 2015, he co-wrote with LEST-CNRS the *Guide de l'Innovation Sociale: comprendre, caractériser et développer l'innovation sociale en PACA*. He is also an administrator of ENERCOOP PACA, which is an SCIC-SA (Société Coopérative d'intérêt collectif-société anonyme).

Megan A. Haddock is International Research Projects Manager at the Johns Hopkins Center for Civil Society Studies. She is responsible for managing the strategic design and implementation of an ambitious array of international projects that seek to improve the current understanding of the non-profit sector and of volunteering worldwide. Ms. Haddock is a lead author of the International Labor Organization's *Manual on the Measurement of Volunteer Work*, 2011, the first official guidance system for the development of data on the size, role and character of volunteer work at the national level by government statistical agencies. Ms. Haddock received a master's degree in public policy from the Johns Hopkins Institute of Policy Studies and a bachelors' degree in international relations and political science from Carleton College.

Akira Kurimoto is professor at the Institute for Solidarity-based Society at Hosei University, Tokyo and director of the Consumer Co-operative Institute of Japan. He served as Chair of the ICA (International Co-operative Alliance) Research Committee (2001-2005). He is the current vice chair of the ICA Asian Research Committee and member of the ICA Principles Committee. His publications include: "Japan's Consumer Cooperative Movement: A Comparative Review," in *Robert Owen and the World of Cooperation* (1992); "Co-operation in Health and Social Care: Its Role in Building Communities," in *Social Capital in Asian Sustainable Development Management* (2003); "The Institutional Change and Consumer Co-operation: Japanese vs European Models," in *Consumerism versus Capitalism?* (2005); "Peace and Co-operation: Reflections on the Endeavors of the Japanese Consumer Co-operatives," in *Co-operatives and Pursuit of Peace* (2007); "L'action coopérative pour la santé," in *L'économie sociale, une alternative planétaire* (2007); "Structure and Governance of Networks: Cases of Franchising and Co-operative Chains," in *Strategy and Governance of Networks* (2008); "Evaluation of Co-operative Performance and Specificities in Japan," in *The Worth of the Social Economy* (2009); and *Toward Contemporary Co-operative Studies: Perspectives from Japan's Consumer Co-ops* (2010).

Michel Marée is senior researcher at the Center for Social Economy of the HEC-Management School of Université de Liège. He holds a master's degree in economics from Université de Liège and a master's degree in public economics from Université Paris 1 Panthéon-Sorbonne. He is specialized in the conceptual and quantitative evaluation of the social economy, as well as in the analysis of the societal impacts of social enterprises. His research activities focus on social impact assessment. He also teaches business finance, social economy and sustainable development in bachelor's and master's programs at Université de Liège.

Sybille Mertens is associate professor at the HEC-Management School of Université de Liège, where she heads the "Management of Social Enterprises" option of the master's program. She holds a PhD in economics (Université de Liège) and participated until 2004 in the improvement of the statistical evaluation of non-profit organizations in national accounts. In fact, the results of her doctoral research have been utilized by the government of Belgium to – as a first for a country – begin publishing a satellite account for non-profit institutions on an annual basis. She is an academic member of the Center for Social Economy of the HEC-Management School and of the EMES network. She holds the CERA Chair in Social Entrepreneurship at the HEC-Management School and is the author of many scientific publications. Her current research projects focus on the financing of social enterprises, the methods of evaluation of the global performance of social enterprises, the business models of social enterprises and the role of social entrepreneurship in the transformation of the economic system.

José Luis Monzón is professor of applied economics at Universitat de Valencia, president of the Scientific Commission of CIRIEC-España and Vice-president of CIRIEC International, director of *CIRIEC-España, Journal of Public, Social and Cooperative Economy* and member of the editorial boards of several Spanish and international journals. He is the director of both the Spanish and the Ibero-American social economy observatories. He has managed several Spanish and international research projects on social and cooperative economy, professional training and labor market, having collaborated as an expert with the European Commission and with the Social and Economic European Committee. He is the author of numerous publications. Among the latest, stand out *The Social Economy in the European Union* (with Rafael Chaves, last update in 2012) and the European Commission's *Manual for drawing up cooperative and mutual society satellite accounts* (with José Barea). He has directed *The main figures of the social economy in Spain (Las grandes cifras de la economía social en España)*, *Cooperative, mutual society and mutual provident society satellite accounts in Spain 2008 (Las cuentas satélite de las cooperativas, mutuas y mutualidades de previsión social en España. Año 2008)* and *Foundations' satellite accounts in Spain (Las cuentas satélite de las fundaciones en España).* He also directed a report promoting the development of the Social Economy Act to the Spanish Ministry of Employment. He received the first José-Barea Medal of the Spanish Fiscal Studies Institute and CEU San Pablo University in 2014.

Daniel Rault was technical advisor to the Délégation interministérielle à l'innovation, à l'expérimentation sociale et à l'économie sociale (DIISES) in France, whose responsibilities have since been taken over by the State Secretariat in charge of the social economy under the Ministry for the Economy (France). In 1998, he co-authored with Jean-Marie Nivlet *Associations régies par la loi de 1901*, a report on the implication of statistics for associations in France, issued by the Conseil national de l'information statistique (Cnis). He also organized an extensive research program that resulted in the book *Les dynamiques de l'économie sociale et solidaire*, published in 2006 and edited by Jean-Noël Chopart, Guy and Daniel Rault Neyret. He was a member of the Cnis working group, which in 2011 published the report *Connaissances des associations*, signed by Édith Archambault, Jérome Arrardo and Brahim Laouisset.

Damien Rousselière is full professor of economics at AGROCAMPUS OUEST (Public University of Agricultural Sciences in Angers and Rennes) and visiting professor at Université du Québec à Montréal (since 2010). Previously assistant professor at Université de Grenoble, he holds a doctorate in industrial organization from Université de Grenoble and was a postdoctoral fellow at the Canada Research Chair on the Social Economy (2007), where he contributed to the statistical survey on the social economy in Montreal. An external consultant for the Environment Directorate of the OECD, Dr. Rousselière is a specialist of quantitative methods applied to the social economy and in particular to agricultural cooperatives. He has published among others in *Canadian Review of Sociology, International Review of Sociology* and *Journal of Economic Behavior and Organization*. He serves on the scientific councils of Coop de France (French federation of agricultural cooperatives) and Végépolys (a French cluster on horticulture, seeds and urban landscaping) and is a member of the CIRIEC International Scientific Commission on the "Social and Cooperative Economy".

Lester M. Salamon is a professor at Johns Hopkins University, director of the Johns Hopkins Center for Civil Society Studies and serves as scientific director of the International Laboratory on Nonprofit Sector Studies at Moscow's Higher School of Economics. Dr. Salamon pioneered the empirical study of the non-profit sector in the United States and has extended this work to other parts of the world. Author and editor of more than 20 books, his works on the U.S. non-profit sector include: *America's Nonprofit Sector: A Primer*, now a standard text used in college-level courses on the U.S. non-profit sector; *The State of Nonprofit America*, Volume 2 (Brookings Press, 2012); *The Tools of Government: A Guide to the New Governance* (Oxford University Press, 2002); *The Resilient Sector: The State of Nonprofit America* (Brookings Institution Press, 2003); and two volumes that explore emerging forms of philanthropy – *Leverage for Good: An Introduction to the New Frontiers of Philanthropy and Social Investment*, and *New Frontiers of Philanthropy: A Guide to the New Tools and Actors Reshaping Global Philanthropy and Social Investing* (Oxford University Press, 2014). In addition, Salamon has authored hundreds of articles, monographs and chapters that have appeared in a wide variety of publications including *Foreign Affairs, The New York Times, Voluntas, Annals of Public and Cooperative Economics, Stanford Social Innovation Review* and *Nonprofit Quarterly*.

S. Wojciech Sokolowski is senior research associate for the Johns Hopkins Center for Civil Society Studies. Dr. Sokolowski received his PhD in sociology from Rutgers University. He has taught at the Defense Language Institute, Hartnell College, Rutgers University and Morgan State University. Dr. Sokolowski is the author of *Civil Society and the Professions in Eastern Europe: Social Change and Organization in Poland* (Plenum/Kluwer, 2001) and co-author of *Measuring Volunteering: A Practical Toolkit* (Independent Sector/United Nations Volunteers, 2001), *Global Civil Society, volumes 1 and 2* (Kumarian Press, 2004) and the *Manual for Measuring Volunteer Work* (ILO, Geneva, 2011). He has advised national statistical agencies in the United States and abroad on the development of data systems reporting on non-profit institutions. His publications have appeared in the *International Journal of Voluntary and Non-Profit*

Organizations VOLUNTAS; Annals of Public and Cooperative Economics; Nonprofit Management & Leadership; Northwestern Journal of International Human Rights; Journal of Civil Society; International Journal of Contemporary Sociology; International Journal of Cultural Policy, and several edited volumes.

Roger Spear is professor of social entrepreneurship; member of the CIRIEC International Scientific Commission on "Social and Cooperative Economy"; founding member and vice president of the EMES research network on social enterprise; and instructor of organizational systems and research methods at the Department of Innovation and Engineering, Open University (United Kingdom). His research on innovation and development in the third sector, particularly social enterprises, includes: a study of work integration in several European countries; two comparative studies of social enterprises in Europe (see: www.emes.net); and a major study on governance and social enterprise. He was one of the coordinators of a major CIRIEC project on employment and the third system (funded by the European Commission). Other projects include an evaluation of the UK government's social enterprise strategy; projects on social enterprise in Eastern Europe and former Commonwealth of Independent Countries, funded by the United Nations Development Programme; two European Commission projects (one on the social economy from the perspective of active inclusion and the other on the French social economy); and three OECD projects on the social economy in Korea, Slovenia and Serbia. Other policy work includes a World Bank project in Egypt on social entrepreneurship and a mapping of social enterprises in Europe. He is currently guest professor at Roskilde University, Copenhagen, Denmark, contributing to establishing an "International Master's Degree in Social Entrepreneurship."

Martin St-Denis holds a master's degree in economics from Université du Québec à Montréal (2014). He has worked as a research assistant at the Canada Research Chair on the Social Economy (2010-2013) and for the Research Group for Human Capital (2013-2014). Part of his research activities focused on conceptual frameworks for producing statistics on the social economy. He also wrote a thesis on the impact of different types of daycare services on early childhood development. He now works as an economist for MCE Conseil, a firm that consults social economy enterprises and worker unions.

Nicoleta Uzea holds a PhD in agricultural economics from the University of Saskatchewan and is currently a post-doctoral research associate with the Ivey Business School, Western University, Canada. Her areas of expertise within the field of cooperatives include: cooperative governance, capital financing in cooperatives, and the economic impact of the cooperative sector. For instance, her critical analysis of economic impact methodologies conducted for the Measuring the Co-operative Difference Research Network (MCDRN) informed the first National Study on the Impact of Co-operatives (NSIC) in Canada. She also served on MCDRN's advisory committee for the NSIC. Her work on cooperatives has been published in the *Research Handbook on Sustainable Co-operative Enterprise: Case Studies of Organizational Resilience in the Co-operative Business Model*, the *Journal of Rural Co-operation* and the *International Food and Agribusiness Management Review.* She is a reviewer for the *Journal of Co-operative Organization and Management.*

Social Economy & Public Economy

The series "Social Economy & Public Economy" gathers books proposing international analytical comparisons of organizations and economic activities oriented towards the service of the general and collective interest: social services, public services, regulation, public enterprises, economic action of territorial entities (regions, local authorities), cooperatives, mutuals, non-profit organizations, etc. In a context of "large transformation", the scientific activity in this field has significantly developed, and the series aims at being a new dissemination and valorization means of this activity using a pluri-disciplinary approach (economics, social sciences, law, political sciences, etc.).

The series is placed under the editorial responsibility of CIRIEC. As an international organization with a scientific aim, CIRIEC undertakes and disseminates research on the public, social and cooperative economy. One of its main activities is the coordination of a large international network of researchers active in these fields. Members and non-members of this network are allowed to publish books in the series.

Series titles

www.peterlang.com